# The

# Poetic

# Universe

# THE POETIC UNIVERSE

## We Are It

Appearance and motion wholly create
Being and time in the arena of space.
We're the complex composites, from simple verse,
Far advanced, perhaps, in the universe.

## Opposites

Religion declares its dogma all at once,
Which is why it comes out as the dunce.

Science is of a repeatable stance,
And thus brings forth reliance.

## And God Created Woman

God offered Adam a perfect version of woman,
One who would even paint ceilings, cut grass,
Work on cars, take out the garbage, and so forth,
But, this would have cost Adam an arm and a leg.

So Adam said,
"What can I get for just a rib?"

## Shakespeare Had the Idea

Such tricks hath strong imagination,
That if it would but apprehend some joy,
It comprehends some bringer of that joy;
Or in the night, imagining some fear,
How easy is a bush supposed a bear!

# The Hand of the Potter Shakes

At the crossroads of His human experiment,
God wondered where our human nature went;

"Damn! I formulated it so perfectly in the lab,
So why did it not turn out exceedingly fab?"

"Adam and Eve failed, in the blink of an eye,
So I sent the Commandments down from the sky,
But ever did humans build the golden calves,
Diminishing my needed adoration into halves."

"So I killed all experiments but Noah's sake,
For sapiens were all a big rainbow of My mistake;
I'll right the human course yet once again, to sail
Into those waters where it can never fail."

"What's this! I see that all has again gone amiss;
I'll have Jesus preach the other check to kiss.
Human nature fails again; they put him to death;
I must check my formulas again, to save the rest;

I'll send more prophets to shake the mixture up;
Oh, of new life cast, they still crash—I give up!

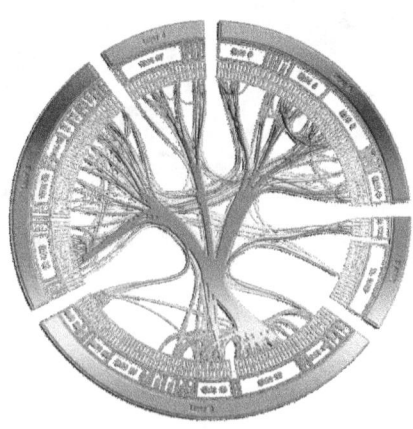

With Earth's first Clay They did the Last Man knead,

And there of the Last Harvest sow'd the Seed:

And the first Morning of Creation wrote

What the Last Dawn of Reckoning shall read

# Duration and Extent

It always get down to
The forever of a passed eternity,
With yet more to come,
After every instant of now.

It doesn't much matter
What the forever Totality state is,
Whether it be the stuff we see,
Or a smaller basis of it,
Or the forever distribution of it
As the sum-thing of nil.

It is that everything possible has
Already happened, and will again.

Add infinity to this, as the extent,
And it is that everything possible is
Also happening somewhere out there
At this very moment,
Even multiple times.

Thus, Totality contains its
Own history and future
At any given moment.

This is all why some call it
'Infinity for Eternity'.

One can see that Totality itself
Could have had no creation—
And thus there is no Creator.

These proposed Great Minds

Or Super-Computer-Projectors
Would be but secondary, alien things,
Complex and composite.

Not too many thinkers get this far,
Stopping prematurely
At some made-up thing,
Usually something made-up
In our own mammal image,
Or some other crazy,
Non-fundamental scheme.

(We are having fun with da mentals.)

P.S.

Cherish your freedom,
For your time has come,
Within this small parentheses of eternity.

To see a world in a grain of sand,
And a heaven in a wild flower,
Hold infinity in the palm of your hand,
And eternity in an hour.

# The Natural Happenings

*Why these little subatomic things, ask the wise,*
*In such amounts and of their special size?*

Well, I agree that this shows they had to be made,
As we see from the causeless quantum's shade;

It's been shown by Aspect there's nothing there,
Underneath this necessarily indefinite disorder,
Whence opposite particles become, unbidden,
For causes beneath causes would have no border.

So, from this causeless bottom, where bucks stop,
Hails the ultimate freedom to live and be a lot.
This scheme, too, hints that the Ultimate Yore,
If it could ever be, would need cause all the more!

Yet, at this very point, which is not an answer,
But a call to think no more, and surrender,
Religion introduces Complexity Infinite
For the downwardly simpler bottom unit.

So, there's no answer given, but only
A larger mystery of the One and Only,
That is an infinitely larger question there,
Rendering the entire 'answer' beyond repair.

While both science and religion claim the causeless,
They are as opposite as could be, none the less,
For one finds no specifics there, none at all,
While the other imagines God's Perfect Ordered All.

If all the above opposed were not bad enough,
There is entirely insufficient evidence for God,

Zero, in fact, in the face of the opposite there,
For the One who is supposed to be everywhere.

Beyond even the total absence of evidence
For the interceding Ruler, an obvious nonpresence,
Leading to the sure evidence of absence,
Is that a first cause can have no reason to it.

Humanists [non theists] push science forward,
God naturally flunking out, with no push backward,
While creationists, with nothing of to push forward,
Ever attempt to push science backward.

This, then, is the end of faith's season,
It being the celebration of rational reason.

# The Elemental Genesis

If the crucial basis came from somewhere,
Nothing is the only possible lair,
For there are no 'wheres' at all remaining
Of the causeless to be its sustaining.

If the basis is eternal—it must be,
Then regresses can't go eternally—
So it's composed of this new Nothing,
Because Nothing can have no beginning.

Now, it is not that Nothing ever was,
For it cannot be and it never does,
But, that nonexistence ever fluctuates,
Both into and out of the two states.

Of some of these waverings of its form,
The opposite particle pairs are born,
For this is all of the natural norm,
As must be concluded by AustinTorn.

Swift as existence hastening to its task,
Of positive/negative, substance springs forth,
Rejoicing in its splendour, and the mask
Of darkness falls from the awakened 'verse.

On Earth, those crystals of the mountain snows
Melted, above crimson clouds, and from the glows
Of the flame, the Ocean's horizon arose,
With flowers in fields or forests which unclose
Their growing vision to the kiss of day,
Swinging their censers in the element
Of eastern incense lit by the new ray—
Burning slow and inconsummably, and sent

Their odorous scents up to the willing air;
And, in succession due, did continent,
Island, sea, and all things that in them wear
The face and complexion of mortal flair,
Rise, as the sun their father rose, of old,
In portions of the soil, which light did mold,
As its own, and then imposed, untold:
All their thoughts that must ever unfold.

## Totality

Totality is necessarily
Causeless,
Being eternal,
And everywhere,
Being infinite;
Else it would
Only be a subset
Or an infinite regress.

It's here,
Therefore it had to be.

Being causeless,
It could not have had
Any definite design
Imposed on it.

Material stuff may not be
Forever durable,
So that may not be
The bottommost base of it,
Plus, how could some eternal stuff
Be already made without
Ever having been made?

An absolute nothingness
Cannot be,
For if it was,
It still would be.

So, there must be
Something instead of nothing.

The two 'impossible' choices
Are forever stuff
Versus
Ongoing stuff as
A distribution of nothing,
For nothing cannot be (stay),
This perhaps being an unstable state.

Yet, one of the 'impossible' choices
Must be true, perhaps even both,
As we sill see;
So, the answers are at hand,
Meaning that we must have
One or the other or both'
No way out.

We see the simpler
Graduating upwards
To the more complex,
So that is a clue.

God and Brahman solutions
Are shortsighted,
Being merely the formation
Of a larger question that
One then takes as an 'answer';
Besides, systems of mind
Doing planning, thinking, and doing
Cannot be first, for the parts
Beneath would have to be there earlier.

The simpler that stuff is,
The more unstable it is,
Excepting the inert,
Ever going through phase changes

And/or recombining upward,
So, thus is a clue.

Energy must be conserved,
Not being able to come from nowhere
Unless it is balanced,
Positive and negative,
So, this is a clue.

The universe is here,
This being any old time and place
That is not special,
So, this is a clue.

M-theory suggests 10^100
Possible solutions for universes,
So this is a clue,
But just maybe,
For who knows
If string theory is a clue.

Quantum theory shows
The emission of
Opposite 'virtual' pairs
Of particles,
Some of which may become
Somewhat enduringly real,
So this is a clue,
Although we don't know
If all particles decay,
Some over very long periods.

Quantum theory shows
That randomness may be
The bedrock of reality.

On the other hand,
Conservation would say
That a particle cannot
Just disappear and reappear
Centimeters from where it was
Without any accounted cause and effect.

So far, it seems true that
While the bottommost basis
Must be causelessly random,
From there on up there
Seems to be cause and effect:
Protons forming stars.
Stars emitting the higher elements,
Some of these atomic elements
Forming molecules,
Which then form cells,
Life, and consciousness,
For, while the initial conditions
May have been arbitrary,
There was necessarily order
From there and then.

Jumping to a conclusion
From the two 'impossible' choices,
It could be that,
Since forever stuff just is,
That it had no prime mover;
However, if it is a distribution
Of the perfectly unstable
And simplest state: nothing,
Then that explains
The ongoing production
Of opposite virtual pairs,

And, since this has been
Going on forever,
It is pretty much the same
As saying that stuff was forever,
So, we can even combine the two;
However, if not it doesn't
Even matter for the conclusions,
These being that we are free to be,
And that Totality had no creation,
Thus leaving out a Creator,
Which is why we are free,
Not to again mention
The impossibility of the
First cause already being complicated.

So, the simple TOE,
As simple it must be,
Is that any old result
Could have come about,
Here or anywhere,
And perhaps always did
And does, any time anywhere.

Either that
Or there is just one universe,
It itself matching Totality,
That recycles itself
Over billions of years,
But this is but a small point,
The important points being
That the notion of the causeless
Is itself the TOE,
This causeless necessarily
Being the law of no law,
One with no particular formation,

And that there could be
No built-in meaning,
Which may be disturbing
To some who wanted
To be watched over,
But, they forget that
They then would not
Have the ultimate freedom
To make their own meaning
Of out life and the situation
Of the human condition.

Note that all is the same result
For those who say that
They don't know.

So, the temporary conclusion is
That the why is
That something must be
Because nothing cannot stay,
The How being some kind
Of balance of opposites
Occurring from the only
Possible prime mover
That is ever infinite and eternal,
Nothing,
Such as the virtual pairs,
Or positive stuff vs. negative gravity,
Or even the polarity of charge
Nullifying all existence,
In the overview,
The ever production of stuff ongoing,
This being akin to it
Having always been there.

It's not as interesting
As we thought it would be,
But, then again,
How could it be,
Way, way down?

The real excitement
Is at the other end of the spectrum:
The complex, where we are,
As thrust into existence,
Which even precedes
In importance that of essence!

(Get a life!)

P.S.

Of course,
This doesn't stop some mammals
From making things up,
Such as that life
Can only come from Life,
Even then abruptly flip-flopping
To then say that the Life
Then didn't need to come from LIFE, etc.,
Their initial basis now in a shambles,
But whoever said that mammals
Had to be sensible?

Some are so silly as to say
That all is fake,
As like for show in a dream,
Only then to have some
Actually really real Guy beneath,

Named Brahman,
Pops up all of the sudden
Drowsing, sleeping,
And dreaming his life away,
His own means of being
Totally unexplained and ignored,
This being some kind of strange
Psychological stopping point
Of complete satisfaction,
Totally contrary to their
Big need for explanation
Of a lesser question in the first place.

Well, some mammals
Are not just vainglorious
And far from humble,
Thinking themselves
To be the center of all
And the measure of all things,
But just plain weird,
Making outright baseless
And ungrounded
Pronunciations by fiat,
Even then deceptively
Preaching these theories
As fact and truth
To the young and unsuspecting,
Even compounding that lack of ethics
But trying to protect the beliefs
Of the imaginary invisibles
By going to war over them,
Onto their own bloody end!

# A Letter from the Futurians

And it came to pass that God, Adam, Noah, Moses,
The Ten Commandments, the prophets, Jesus,
And Our Lady of Fatima were of no real help
In reducing the sins of the horrible human nature
Invented for man by God Himself;

And so it came to pass even more, not less,
That it all really appeared hopeless
For the higher mammal species known as Mankind
(Women were secondary to God/Religion),
As these homo sapiens had not
Even remained at the same level,
But had gotten even worse,
Some of this evil even being performed
In the direct name of God.

It also always comes to pass that God
Never gives up on his failures of creating
A temptation-resistant species;
So, a new search is now underway
For a teen-age virgin, engaged to a saint,
For GOD to impregnate,
With his daughter to be
(Women are equal now),
Because, as ever, GOD still
Has no wife or girlfriend—
And is not even seeing anyone,
For how could a relationship between
Mr. Right and Ms. Perfect ever work out?

However, so far,
No teen-age virgins have been found
Anywhere on the planet, much less engaged ones,

Princess Diana having been the last of the breed.

Furthermore, the virgin's husband-to-be
Must be saintly enough to let her fool around
With a Ghostly Superbeing for a one night stand,
Although, admittedly,
The Devil would probably
Be much more passionate.

So it came to pass that all of this really needed
To come to pass or humankind would be doomed.

Hail "Mary", full of Joe (hollow be his game),
Wherefore art thee? For the Lord,
Art, in Heaven,
Harold be His name on Earth,
Wants to be with thee, blessed be thy womb,
For it is Heaven's Kingdom!

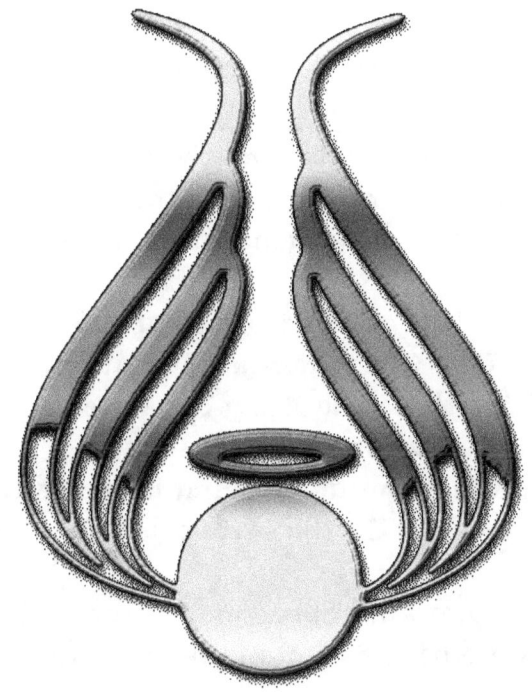

A Useful System
Of opposition-plus-transition
Operating at various levels of the universe,
This being the first one of the 'pyramids':

forces

Strong vs. Weak
opposition
plus the
Magnetic<—>Electric
transition.

The Strong promote stability;
The Weak promotes changeability.

Their balance promotes progress,
For events have to be orderly enough
To take form,
But not so much frozen
That they cannot change.

The Magnetic and the Electric
Transition each into the other,
In the self-regenerating wave.

(We're not worried about
The Electroweak unification,
Since those Big Bang days are gone,
But one could still take
The Electroweak as being in opposition
To the Strong.)

Gravity, a kind of secondary force/effect,
Would be the blending of the 4 forces.

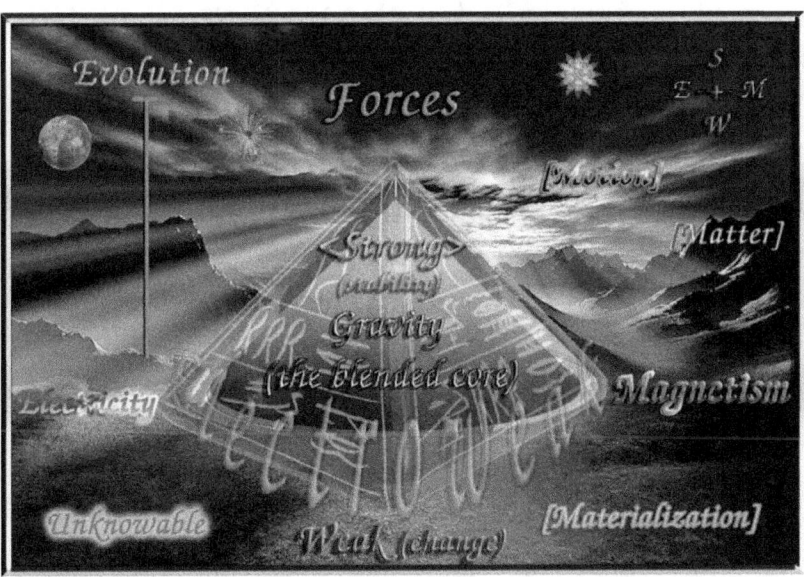

# Being

Space vs. Matter
opposition
plus the
Past—>future
transition.

While space and matter are indeed similar,
The thinness versus the lumps are a contrast,
Granting us a notion of the
'What' in the 'Where'.

Time could go backwards,
But it usually doesn't,
Due to there being so many
More states of disorder than order;
So, for practical purposes,
The transition occurs as
Past —> (now) —> future,
Granting us the passing of
The 'Then' into the 'When'
Via the present—now.

Time is more or less energy—
The movement of
The appearances of Matter
Through Space.

The blend of all this is
The essence of life: one's being,
Which is the 'Who'.

The 'Why', I propose,
Is that, since Nothing

Could not be (stay),
Something has to be,
Yet it must amount to Nil,
As there can be no other mover prime.

The 'How' would be
The necessary production of opposites
From the simplest, unstable state: Nothing,
Which could be known as
The opposite virtual pairs emitted,
Or, as like Hawking,
The positive kinetic energy of stuff
Ever being canceled out
By the negative potential energy of gravity.

There is more to this
'Being' opposition-transition,
Which is found by
The further combining the fields,
Such as Space and Past becoming Remembrance,
Matter and Past becoming History,
Space and Future becoming Wishes.
Matter and Future becoming Progress,
With, then, further combining, on up, etc.,
As one may predict...

History and Progression leading
To a Change-in-Structure,
Remembrance and Wishes leading
To a change-in-outlook,
Remembrance and History leading to Learning,
Progress and Wishes leading to Vision,
Onto Direction in life, Creating,
Growth. and Planning.

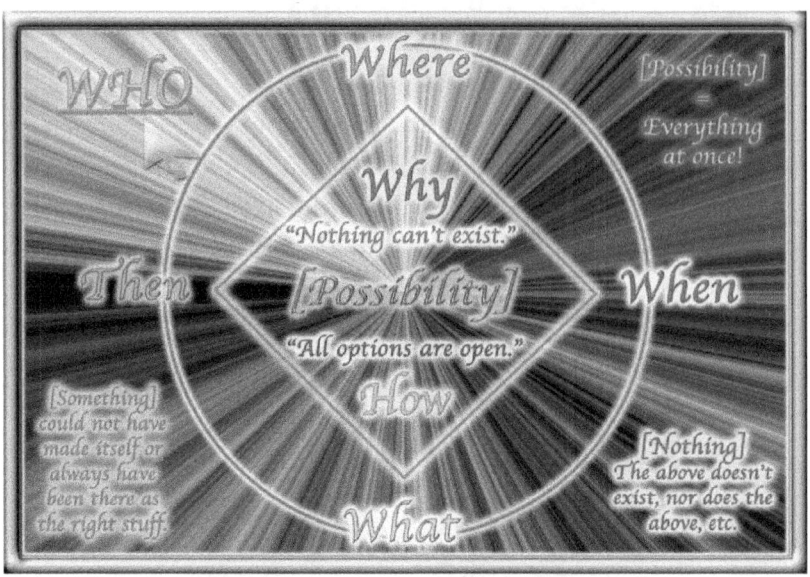

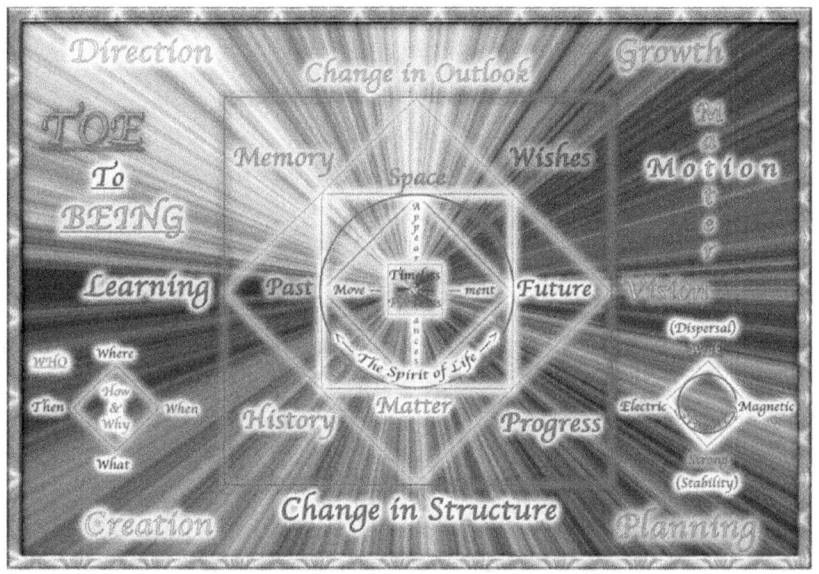

# The Eternal, Infinite Universe

Positive vs. Minus Polarity
opposition
(In the form of charge)
plus the
Mass<—>Energy
transition
and Matter<—>Light
annihilation.

4D space-time
Is like an infinite number of 3D Spaces
Stacked upon one another,
Which I call the 4D Hypercube.

It hard to visualize 4D,
But one can generalize it
from a stacking of infinite 'pancakes'.

So, in seeing this 4D Hypercube
Of 3D Spaces,
One can visualize
That Time is the difference of Space(s),
It serving as both motion, charge, and energy.

And also that Space is
The difference of Time,
This being distance.

Energy/mass is basically curved space.

Space is physical,
But not material.

3D Space is infinite and is therefore
The bounding "surface"
Of the finite 4D Hypercube.

The exterior of infinite largeness
Bounds the infinitely small interior
Of infinite smallness,
An inescapable consequence of closure—
Closure being the level of existence
Where the large
Becomes indistinguishable
From the small,
Both merging into
The singularity of Totality: zero.

Imagine an enormous sphere in space,
Growing progressively more immense.
As long as its volume remains finite,
It is a three-dimensional object
With a two-dimensional surface.

When its volume reaches infinity,
Its three-dimensional interior
Is now the surface by which it is bounded.

The sphere becomes inverted,
Just as infinite smallness
And largeness can be inverted.

The largest, via dispersal,
And the smallest, via compaction,
Are the same vacant ends
Of Totality's spectrum: zippo.

The two halves of Totality,

Delineated by polarity (charge),
Of positive Yang
And negative Yin
Nullify all of existence,
In the overview.

This is because the only
Possible Prime Mover
That is infinite and eternal
Is Nothing,
Requiring nothing prior
To itself but itself.

Yet, it is so unstable
That it cannot absolutely exist,
Even for an instant.

And so it is that we reside here,
In the Balance of Opposites,
Necessarily at the mid-point
Of the largest and smallest infinities.

It is now,
For that's all there can be,
Yet every 'now'
Is out there, somewhere,
The universe containing
Its own history and future,
Even being able to use
The future in the present,
For no-origin systems
Are their own precursors.

The two 3-D quantities
Of 4D hypervolume are

$$distance^3$$
(Space)
And
$$time \cdot distance^2$$
(Energy)

Energy moves through space.

The space of our universe
Is three-dimensional
Because this is
The only dimension
Whose volume is
Compositionally consistent
Through all levels of infinite size
While forming the surface
Of its own hypervolume.

Time is the dimension that bounds,
Not extends,
Three-dimensional space.

Unit hypervolume
Is the internal product
Of time and space,
But it is also the product
Of energy and distance.

The speed of light
Is the dimensional equivalent
Between space and time.

Energy density
Is the 4th-dimensional slope of space.

Just as Planck's constant is
The four-dimensional quantization of photons,
Elementary charge is the four-dimensional
Quantization of particles.

Photons are
The encapsulation of time by space;
Particle fields
Are the encapsulation of space by time.

Note that I have not really mentioned
Anything other than space
Or what could not be an aspect of
Or reduced to space
(Such as its curvature).

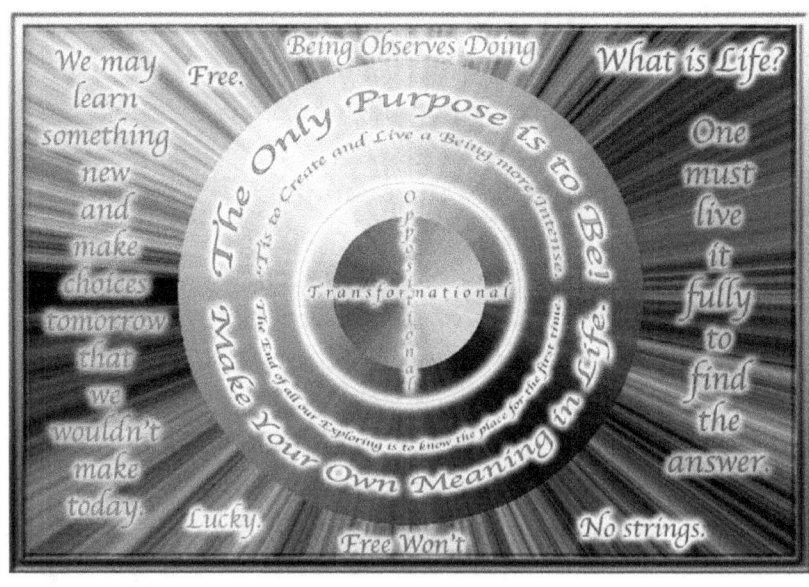

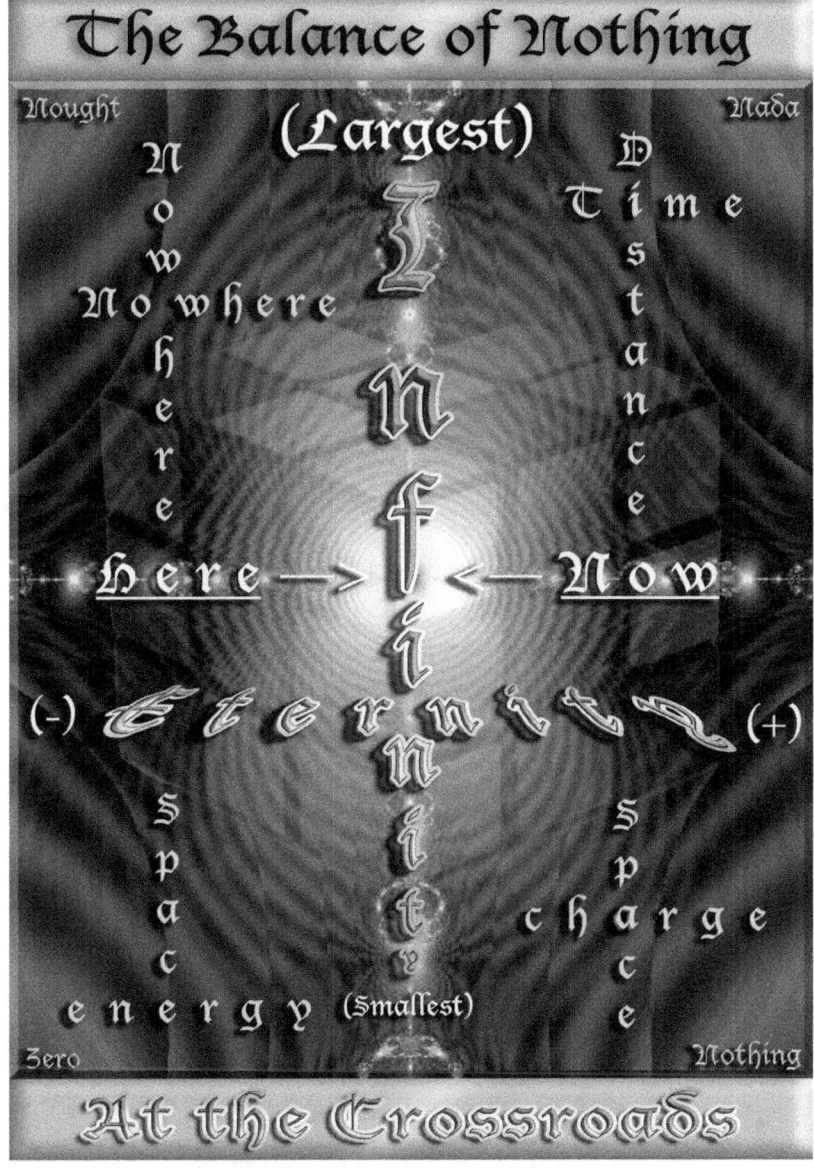

The Balance of Nothing

Nought · Nada

(Largest)

Now · Nowhere

Time · Distance

Infinity

Here → ← Now

(-) Eternity (+)

Space · Space

charge

energy (Smallest)

Zero · Nothing

At the Crossroads

## Stable Existence:

AntiMatter(−) vs. UncleMatter(+)
opposition
plus the
Photons<—>electrons/positrons
transitions
of light<—>matter

There are only two stable matter particles,
The electron(+) and the proton (+),
Because there are only
Two ways to make them.
(A photon is neutral since its has both
A positive and a negative aspect).

## Finite Existence

Largest Infinity vs. Smallest Infinity
opposition
plus the
Past Eternity<—>Future Eternity
transition

Yes, the no-origin Totality
May use its future in the present,
For it is its own precursor.

# Two Investigations
## Of the Only Remaining Big Questions:

## Consciousness and Life

It is best to surround
A problem first,
To localize it,
So that we know where
To look to ask "how".

For example, we know that:
No brain, no consciousness;
Altering the brain alters consciousness;
Anesthesia to the brain prevents
The process of consciousness;
People with a split brain,
Due to the cutting of the corpus callosum,
Such as to relieve epilepsy,
Develop two fairly separate consciousnesses
(The brain stem is still common)
That cause the left and right sides
Of the body to be at odds with each other.

So, consciousness is of the brain, for sure,
But how does "what it is like"
Become of neurological states correlating to it?

It must be that information,
Neurological in this case,
Gives rise to conscious knowledge of it,
Kind of an 'it' (realness and knowing)
From 'bit' (information).
This is where we are led.

What about the indefinite quantum states
That collapse into the real?
This also seems as 'it' form 'bit'.

So, then, using a bit of the preceding,
What about the animate from the inanimate?

Following are the conclusions of the moment
Although, on some days,
I don't admit to the quantum explanations,
But, you know, it seems that something
Must be going on that science
Doesn't yet totally know everything about.


## Life Explained

It is, of course, that atoms make it,
Through a casual nexus
Of physical-chemical reactions;
However, this observation cannot
Be equated to an "explanation",
For it seems not to be reductive,
And so we must delve deeper,
For there may very well be a background
Behind what the chemicals do.

According to the quantum realm,
"Matter" is only composed of potentiality—
It only becomes matter when it's "real-ized".

In a stable configuration of matter,
Such as in the inanimate,
All the quantum uncertainties
Are effectively statistically averaged out,

This thus being deterministic;
But in the case of the statically unstable
But dynamically stable configurations,
The "lively" features of the underlying
Quantum structure have a chance to surface
To the macroscopic level; that is, to life.

The electric dipole moment of biomolecules
Might be the ordering parameter
for the corresponding macro-quantum system,
And so this results in a change in quality
for that macro configuration.

There is the particle and there is the wave—
Either one forced on us by our observations,
Being jointly known as the 'wavicle',
All three states of which are not the actual reality.

There are, strictly speaking, no objects
That are identical with themselves over time—
The temporal sequence remains open;
Nature is no longer seen as clockwork,
But only as a "possibility gestalt",
The world occurring anew each moment.

The deeper reality from which the world arises,
In each case, acts as a unity
In the sense of an indivisible "potentiality",
Which can realize itself in many possible ways,
It not being a strict sum of the partial states.

What remains unchanged over time
Are certain properties that find expression
In the laws of conservation of energy,
Momentum, electrical charge, etc.

It appears to us, though,
That the world consists of parts
That have continued from "a moment ago",
And thus still retain their identity in time;
Yet, matter really only appears secondarily
As a congealed potentiality,
A congealed gestalt, as it were.

Physical phenomena are not made
Of basic building blocks
But are made of "elementary processors",
Which are complex-valued field "operators"
That depend on time and location.

These generate certain overlappings
Of correlated multi-dimensional wave fields
That are propagating through time,
Fields of possibility,
Whose intensity is a measure
Of the probability of
An object-like realization,
This intensity being very sensitive
To the relative phase
Of the overlapping partial waves.

There are no point masses,
But only smudged particles,
Such as we know of
In the space-filling representations
Of the distribution of electrons
In the shells of atoms—the 'cloud'.

There is a relationship structure
That arises not only from the manifold

And the complicated interactions
Of the imagined building blocks of matter,
But also one that is substantially more inherent
And holistic, again such as
We see in quantum physics.

So, there is form before substance,
Relationality before materiality.

It's hard to imagine
Pure relationships existing
Without a material substrate,
But, consider electromagnetism:
It fills space—without a material substrate,
Or consider a music CD—
Its singers and instruments encoded
In a relationship structure.

The material CD is only a carrier,
Of secondary importance,
Its information being primary,
An analogy to particles
And waves' descriptions.

Impressions of realizations
Are left in our 3-D world by the gestalt
That "lives" in the multi-dimensional spaces
Of quantum superpositional possibility.

Quantum systems
Of many quantum states
Are not so much systems
As they are holistically differentiated
Process structures.

However, considering them as systems,
They are complex, meaning here
That such systems cannot be reduced
To simpler systems
Without breaking connections;
Thus there can be no reductions,
For, as in chaos theory,
There are embedded instabilities—
And if we disregard even
The tiniest correlations
Then we may severely distort the result.

We can no longer just analyze the parts
But must try to use much more
Sophisticated statistical methods,
These being more than the simple probability
To which we are accustomed.

Waves can reinforce,
Weaken, or even cancel out,
This all being a kind of generation

Of partial disconnectedness
By intermediate extinctions,
Such as in the way
A biological organism forms
From a single cell
By successive cell divisions,
Which do not occur by parting,
But by repeated formation of
Semi-separating cells walls;
However, this is only a very rough analogy.

Via metabolism,
Life forms have a

Sufficiently powerful energy pump,
One that could conceivably generate states
Of thermal disequilibirium
In molecular systems
Embedded in certain substrates
That would excite certain
Low-frequency collective modes
Of vibration with great power,
Perhaps via mechanisms similar
To Bose-Einstein condensation,
The electric dipoles coming into play
As an ordering parameter;
However, this is not
A conclusive, direct connection.

Information appears only in the animate,
And is furthermore exchanged,
The meanings somehow combining
To make sense in
Some nonreductive process—
The relational reality of life happening
At this semantical level
Of information exchange.

Life is not mindless; it is inspired;
It's meanings cannot be
Discovered by observation,
But only by participation.

Life's entities embrace one another:
Cell, organism, species, and biotope.

A living creature is more like a poem,
Revealing further dimensions
And expressing new properties

At every level of organization:
Letter, word, sentence and [uni]verse.

Somehow, perhaps, quantum states
That continue on further
In the quantum superposition
Have reached more efficiency and effectiveness,
For all paths are tried out,
Just as in the 95% efficient
Photosynthesis methods seen,
And so that's what collapses out of it,
The more productive paths that last,
One usually with the least amount of effort, too.

## Biological Consciousness

As for the theme that consciousness
Is made by the brain,
This is confirmed again since the results
Appear 'in it' 200-300 milliseconds
After the brain has finished its analysis,
For this takes some time, as we would imagine.

Perhaps a snail has
A limited smudge of consciousness
In which thoughts of just warmth and cold,
Light and darkness, surface.

We know that consciousness is of the brain
Because we can introduce
Molecules into brain areas,
This being anesthesia,
And turn off consciousness completely;
Consciousness has no independent existence.

Whatever it is that the brain does
To achieve consciousness
Can be stopped by anesthesia when it dissolves
In the oily regions of the neuron microtubules.

The brain then stays active,
But it does not produce any consciousness
Until the anesthesia is taken away.

The same kind of result occurs when you faint;
Consciousness is therefore surely of the brain;
As such, consciousness can be turned off and on
By the xenon or isoflurane gas of anesthetics.

Our environment, inside and out,
Is symbolically represented in the brain,
Our memories and cross-associations
Recognizing and remembering
The meaning of what we 'see', think, feel,
And witness in the unified experience
Of living life as a being.

We all know how wonderful and quick the brain is,
It nearly instantly processing visuals, sounds,
Touches, tastes, and odours
On into higher and deeper systems;

It searches memory very quickly
For what is known,
Such as what the letters
And words of this essay mean,
Forming an abundance of
Further thoughts and actions
Based thereupon, and so forth, and so on,

One hundred billions brain cells
Winking and blinking and connecting,
Making their results known consciously,
At the last, continuing on
In a train of thoughts ever becoming
And arising in almost a kind
Of competition for attention.

Brain cells (neurons) have
A hundred billion connections among them,
Their 'firing' depending on their inputs.

Electricity carries the 'message'
Through the length of the cell
To the gap (synapse),
Where the message turns
To chemical (neurotransmitters),
To take it to another neuron,
Wherein it becomes electrical again,
And so forth.

Your brain neurons have been arranging
Their connections all your life;
It is what you have become,
Molded by your experience and learning.

You are a bio-electrical-chemical being.

# Consciousness and the Brain

The brain is amazing and can sort
100 million bits of information in an instant.

Consciousness must play an active role
In the functioning of the brain,
For both evolved together;
They are intertwined in a process.

Consciousness, being a global physical,
Must still yet tie into the separated physical;
So, how is consciousness mediated?

It must be at the synaptic cleft
That what we call 'mind'
Qualitatively meets the brain proper.

Here is where
The neuron fires or not;
Here is where a neuron
Meets other neurons;
This must be where data
Turns into thought;
Whatever triggers
These switches
Produces thought
In consciousness.

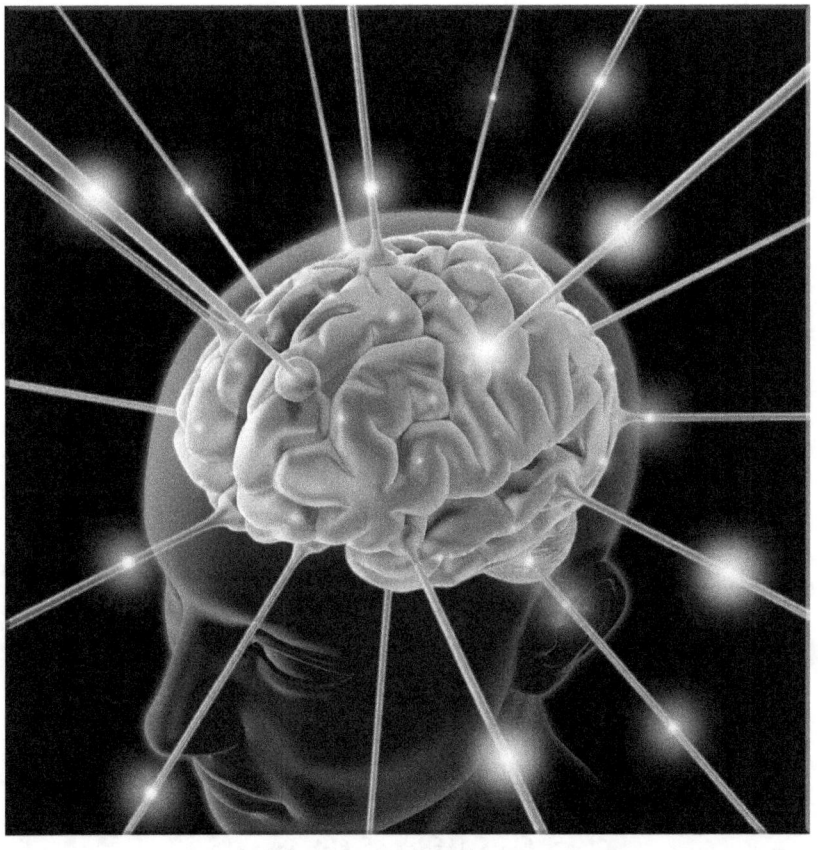

## Acceptance

I accept almost everything,
Except maybe mosquitos,
Although they are probably
A necessary part of the food chain,
Because the world is perfect
Just the way it is
for the evolutionary stage it is at.

Of course, any one thing on Earth
Probably has a mixture
Of good and bad uses,
But, if we were to take it away,
The whole balance of the earth might collapse

Yes, we have crazy emotions sometimes,
Some of which are rather useless now,
Or almost so,
That are still forced upon us,
Very much uninvited,
Such as jealousy and anger,
And, too, children's emotions are not mature
And teenagers have wild hormones
And holes in their brains,
And both apparently
Bring great suffering to all,
Not to mention what the deviants
And low-lifes of the world
Do to us and each other—
But all this is just a stage
That the very primitive
And mostly infantile human race
Is floundering through.
I accept even death,

For without death I wouldn't even be here,
For my human ancestors evolved because of it
And if they would have lived forever
Then not much DNA would have ever changed.

I accept the natural world
And even the humans mentioned above
That are just as much a part of natural world
As anything, crazy as they are,
For, without the beast in us, for example,
There would have been no zest for life.

We have probably survived
Not in spite of being brutal,
But because of it, and at least it may have led
To cooperation for the hunt and for war.

I accept bacteria
And all the diseases they bring,
For they also ferment the soil
And aid digestion.

I accept bugs and worms,
Although I am still working on
Accepting mosquitos.

Worms, for example,
Aerate 400 tons of soil per day.
No worms, no life;
So, please give a worm a hug today.

The same with flood and drought,
For the universe has our well being at stake
Only in the most general sense,
But not in the specific sense.

A farmer's crops may dry up
Or get washed out, but, again,
If there is no water, then there is no life.

Well, anyway, it's warm here,
But there are no mosquitos,
So, I'm really extra happy about that.

## Progress

Astrology gave way to astronomy,
Alchemy to chemistry,
And religion to philosophy and science.

Discovering truth provides freedom
From the shackles of myth; it is not doom,
As the notion of Pandora's box
Deceptively paints it.

The box of truth opens by itself,
No matter how much
One might try to put a lid on it.

## Not Progress

It seems that some may have
Such a wide open mind
That emotions such as hope and despair
Can trump the evidence of the senses,
Especially in times like these,
When the world seems to
Be spinning out of control,
When there is a surge of belief in astrology,
ESP, and other paranormal phenomena,
Spurred in part by a yearning
To feel a sense of control.

The mind looks a bit too hard for explanations,
Especially for those that let one be connected
To a larger reality, a comfort bolstered
By the thought of angels who are watching over one.

Yet none of this is ever so clear cut
That one can ever see it straight out.

## Dead Ideas

Strewn about this great panoramic realm
Of the ONE possibly conceivable at the helm
Are all of the unknowable fabrications
Often dreamt up via exaggerations
By the human race of mammal sapiens.

The realm of such pronouncements has come to be
Superposed at the furthest edge of reality,
Poised by the scope of some wishful thinking,
By all those dreaming and wild supposing,
Who wish for such legends to be ever
Actualized and realized; however,
These ideas have never ever made it
Into our observable realistic habitat in any way,
They but remaining in the minds, joint,
Of the idea-beholders—
Even as widely varying viewpoints.

Without so much much as a word to say,
I passed those to whom most no longer pray,
Nor believe in, but once did, namely,
Those of the graveyard tombstones now so unholy:

Astrology—the God of the Stars that plod,
Eternally blazed and marbled in the sod,
The monuments of Diana the Moon God,
Apollo the Sun God , Baal, Zeus, Wotan,
Aphrodite, Thor, Mithras, Isis, Amon,
Poseidon, the Druid Gods, and on and on, anon.

I ever hurried past the ledgering
Of those older mythologies preceding
The formation of the Old Testament story—

Those ancient superstitions whose very
And various olden amalgamations
Came forth to form it whole for our salvation.

I paused at that Old Testament maligned,
To mark the old but lingering lines
Of the 'knowing' of more invisibles,
The beliefs in imagined angelic creatures...

There were Angels standing, frozen in stone,
Over the timeworn memorials' poems,
And atop the crumbling gateposts,
They being the livelier and near-living ghosts
Of the representations of the three spheres
Of the heavenly host: the demigod-near
Seraphims, Cherubims, Ophanims,
Thrones, Principalities, Dominions,
Powers, Archangels, Angels, and, those final,
And perhaps the most useful—the Guardian Angels
That are said to protect children from falls.

There, Amaranth, its dead red leaves never
Ever fading on this Earth unto forever,
Gave some color around the graveyard pallor
And to the dateless headstones' squalor.

There's a garish purplish maroon view, on high,
Of streaking lights of an electromagnetic sky,
Heretofore never sighted by my self;
I strolled on and into the vale itself.

## First Universal Equation
## Of Singularity

Largest x Smallest = Finite Unity (1)

as in

Infinity x Zero = 1

Thus

$1/0$ = Infinity

and

$1/$Infinity = 0


## Second Universal Equation
## Of Singularity

Largest x Smallest = Finite Unity (1)

as in

Infinity x Infinitesimal (Point) = 1

Thus

$1/$Infinitesimal = Infinity

and

$1/$Infinity = Infinitesimal

# Three States

A photon can become an electron and a positron,
Revealing the photon's peaceful neutral aspect
Of plus and minus residing together;
An electron and a positron can
Then recombine into a photon.
This is a fine stability.

In fact, all types of matters and their antimatters
Produce photons (light) when annihilating.

An electron and a proton reconcile
By making a hydrogen atom.
Perhaps a proton may, after a really long time,
Decay into a positron, again leading to light.

Three stable states:
electron(–), proton(+),
And photon(–/+ as neutral).

Polarity (charge) seems to be
The key to our reality.

# Dice Game

We've already proved
That dice rolled,
With the no-beginning coming
From the eternal causeless realm
And that the mass density
Of the universe
Sums to the near 'nothing'
Of the quantum tunneling—
For which the proof matches
Out though experiment,
That whatever had no cause
Could thus have no design
And so there had to be
An indeterminate chaos,
As seen.

So, since God cannot be,
There is nothing that He did;
That's the beauty of the proof—
It cuts Him off at the source.

Then, just for fun,
And to give every chance,
We looked everywhere else,
Finding only the natural—
And no supernatural.

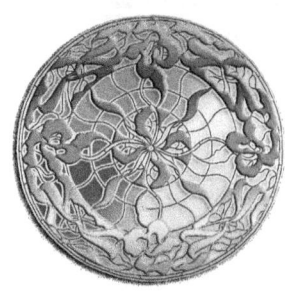

## Night Spooks

We went to an old country dairy
The other day
To get our truck's wiring fixed
By one who knew trucks,
Dropping it near midnight for the next day.

There was a barn and a silo
In the background, kind of a spooky sight.

We heard some voices coming from
Behind an ambulance parked near the garage
And so we looked on one side;
The voices were bouncing off the wall,
So we went around the other side;
A man was laying on the ground,
As if he'd fallen out of the ambulance,
But was actually fixing one of its wheels
And talking to his girlfriend on his cell phone.

We left a note about the truck,
And wandered towards the barn,
Greeted by some chickens and a cat.

There was a ladder outside the barn,
Leading to the hayloft opening.

We went up and in.

# DNA

Proto-man gave way to near-man
And thence to us, eventually,
When two 'monkey' chromosomes fused together,
Making 'us' incompatible with the other chimps

We have 23 chromosomes.
Two of them merged when we were monkey,
As we used to have 24, like the chimps;
This allowed us to go our own way,
For then there was no more monkeying around.
And so our ancestors, then,
Truly descended from the trees!

We came to need no specialized niches,
Since we could adapt to any terrain,
Having brains that could learn much more
After birth than instinct could bestow before.

DNA remembers every step of our evolution—
And you can see this in 'fast' motion
When embryos form, simply, in the liquid womb,
Replicate, and then grow cells
That diversify into a human being
After going through the nonhuman stages.

Thus, four billion years compresses into
The nine months of pregnancy.

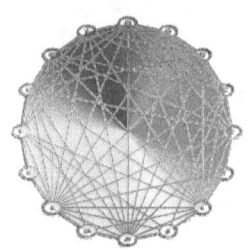

# Non-Centric

It was said,
Probably by the right-brain,
That humans are
The center of the universe...

This is called human-centric,
The stance that man
Is the measure of all things.

It is neither meek nor humble
To suppose that humans
Are so special and/or even deserve
Reward in an afterlife.

The time of our universe happening
Is not of any special moment;
It's here now, at any old time.

Therefore, any universes could be,
Any time, or even at the same time,
And likely anywhere.

And so could be life,
Since for decades now,
Prebiotic molecules
Have been found in space
And on meteors;
So, we're not even likely life-centric.

There is no Earth-centric
In the solar system;
It is sun-centric;
Nor time-centric, as shown,

Nor galactic-centric,
For we are out on an arm,
Nor the Super-Cluster-centric,
Nor even universe-centric,
Nor not likely even
Multi-verse centric.

Pride: Ego exaggerates self-importance
To claim that we're specially created,
Deserving the reward of a divine destiny.

Humility: We're electro-bio-chemical,
Just as organic as anything in nature.

It is pride, perhaps, when one wants
To be King or Queen of reality.

# Poor Observations

It's not that people observe totally differently,
It's that some people outright neglect
What they don't wish for, not even taking it in—
Not doing any observing of fact at all,
For the lone right brain wants what it wants,
Turning some of its neural sensations
Into bare naked claims of magic.

They might
If they want their imaginary notion bad enough—
Even go against all of evolution's evidence.

To do that,
They might even go against all of science.
Going further, some might even say
That the brain does nothing.

Yet, they present nothing but words,
Saying that, if pressed,
That evidence is not even possible or conceivable,
All the while pushing against science and facts
By the same and obvious ploy
Of calling them 'opinions' or 'beliefs',
Mostly because they have no real way
To push forward with their invisible imaginaries.

They'll even get mad if one brings this
Or the invisibleness to their attention,
Although, as we know, anger has no brains.

They think that getting mad
Shows how right they must be.

They are not driven by facts,
But just by what they want.

They might even do a total reverse,
Grasping onto some 'science'
That hints of what they wish for,
But this usually turns out to be 'pseudo science'.

Such is the psychology of the human condition.
Reality cannot made out of wishes.

Science, on the other hand,
Is not just half of one person's brain (the right),
But is the real information and research
Of people who have found things out
That can be repeated and shown, again and again.

They don't use magic power;
They use confirmed information,
Such as that molecule signatures
Can be recorded on Earth at low temperatures;
Then, when they observe space through telescopes,
They can match the molecules' signatures, say,
Then obtain information that there
Are prebiotic molecules out there
That could lead to life.
Same with fossils and DNA.

Then there always comes along
Some lone right brain proclaiming magic powers
To say that, for example,
God is sending me some golden plates
With which to extend the Bible;
Of course then, no one is allowed to see the plates,
Plus, then they inexplicably get 'stolen'.

# Ungrounded Beliefs Upheld

A naked belief hangs in the air,
Because it is but upheld
By the owner of the belief.

Much better to have
A belief based on fact.

A pyramid of fact stands on the ground
And shows that a single top
Can be held in the air through support,
But the single top can itself not
Stand on the ground without support.

# The 'Hold Up'

A belief is that construct that states
We consider something as true;
But considering and knowing
Are two different words.

One implies holding
Something up as true,
While the other stands
On the ground as being true.

## Natural Selection

Aye, the truth of what now we are is:
We're not made direct, by a Wiz, to take a quiz,
But as organic, then mammal, of speciation—
One passing narcissism and self-adulation,
On into the bio-electro-chemical organism
That evolved upon a planet near a star, risen
Of, and along the long and winding mindless way
Of slow time, dust, and selection by death—
That sifts the best from the rest: evolution's breath.

DNA remembers every step of our evolution—
And you can see this in 'fast' motion,
When embryos form simply in the liquid womb,
Replicate, and then grow cells
That diversify into a human being,
After going through some nonhuman stages.

Thus, four billion years compresses into
The nine months of pregnancy.

So, then, hail, and good fortune,
fine fellows and ladies,
And welcome all of you
To the Meadows of Heaven—
A high point of being,
Although we are surely
Still in our infancy.

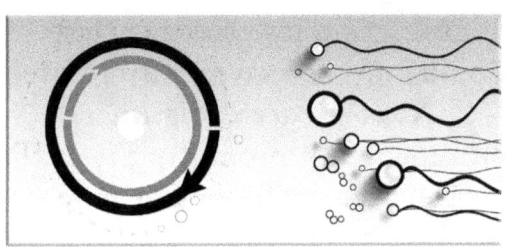

# Debate Mistakes

### 1.

One not having learned the subject matter in depth,
For example, evolution and the proposed method(s)
Of selection, and its history of debate, pro and con,
Some of it concerning Intelligent Design.
Note that Earth Science is necessary, as well,
And not just of current times,
But of the previous 4 billion years.

### 2.

Same as (1), but the case where one
Doesn't study the subject on purpose,
Saying that they have no need for it,
Perhaps due to some 'belief' from a wish.
This is worse than (1), since then they may
Just go against anything and everything.
This will be catastrophic.

### 3.

Not following the debate.
While this may work as a good "pop-in" response
To an isolated point that is very complete,
Chances are that it isn't, usually,
For either it builds upon some prior points
Or the information has been much
Elaborated on elsewhere in great detail.

### 4.

Inserting a pet peeve that
Could only be marginally related,
At some great distance,
But then repeatedly promoting it in the main.
Debaters will wonder why.

5.
Begging the question.
Example: Brahman did it."

6.
Completely changing the topic to another,
In order to cover the embarrassment
That one are not making any headway
On the current topic.

7.
Other avoidance diversions,
Such as name-calling or huffy-puffy bad moods

8.
Rambling.
This usually carries zero information content.

9.
Pushing against all because
One has nothing to push forward with.
Always totally dismantle all the points
That you are against by showing just as much,
If not more, evidence against
And also for your alternative stance.
Else, it goes nowhere.

10.
Don't go bonkers, for you may feel ill from it.
This may even retard your life
At large and your studying.
Also, people may avoid you,
Because of some of the involuntary ills
Imposed on them, which is not a right.

# Cosmic Non-fingerprints

So far, we have seen that
There is no fingerprint of God anywhere;
Plus, theists have not shown that
There is any fingerprint whatsoever.

It really doesn't go anywhere for someone
To just hope that there might be a fingerprint
Or that there will be one.

Where is it then?

Hopes and predictions are not facts.
No moon God; no Mt. Olympus Gods;
No sun God; and now no universe God.

This is not terrible;
It's just that we search for the truth.

Truth is what can ultimately help
The human race, not myth.
The thing, though, about the theists saying
There should be fingerprints of evidence all over,
Completely verifies the approach
Of the disproof of God that finds none.

Note that only one fingerprint would show God;
One measly one, that's all, but, still, there are none!

It is really that humanity is fine-tuned
To the earth and its part of the universe;
Evolution by natural selection has shown us that.

The IDers of the past have been confined

To looking for gaps, but now the gaps close,
To squeeze them out of their last refuge.

Behe, for example,
Just put out the same content
In a new book,
Stuff was already discredited
In his previous book,
He not even including
The new facts thereafter.

For one glorious mishap of his a while back,
Google Dover, Creationism, court trial (Delaware).

In dealing with God, there can be nothing visual,
But the whole is still wished for in the mind,
As should be the implicated details,
But only if we allow them to have a place,
For it is easy to neglect those.

Some might just remain
At a level of "all is one",
Forever placing themselves
Far away from any detail,
Even to the point of not
Being able to consider them.

Furthermore, the mind,
In the case of the imagination
Of the idea of God,
Contains no real and actual details
To guide it for any armchair analysis,
Yet, more layers are even layered on.

To be complete in examining a concept,

The brain must be able to think
Both scientifically and holistically,
Although these views
May not present themselves
At the exact same time;
One needs to juggle them, perhaps,
If the holistic is dominating;
But, just remember that one
Is already one step lost
When one even begins
To imagine invisible things.

The problem with all Designer arguments
Is their failure to account for the nature
And prior existence of the Designer.

Under what set of physical laws did God operate
Before He made the universe,
And where did those laws come from?

We are left in the same Ark as before,
And so we cannot just have the right brain
Exempt God from these real concerns.

If He was Intelligent to enough to engage
In large-scale cosmic engineering,
By what means did He evolve that intelligence'
Except, that he's supposed to be fundamental?

How did He know that the system of natural laws
He chose before the Big Bang would inevitably lead
To thinking creatures like ourselves?

And why did it take so long
If the answer is that He is all powerful?

And if He existed before time,
Then how did he act and plan and make ready
For the genesis that was really just an experiment
(As if He doesn't really know All in advance)?
It would even take time to invent 'time'.

One cannot just have a great system of mind
Being responsible, without allowing that all systems
Must have components beneath,
For their operations
Of thinking, feeling, and doing.

## Religious Anger

In a galaxy, among trillions,
In the middle of nowhere,
On a planet, among billions
Apparently not needed,
Called Earth,
Out on an arm
Of the Milky Way Spiral,
Some organisms formed
Over five billion years,
Leading to over 30 million species,
Among them some higher mammals,
Such as dolphins, chimps, and humans.

These lumps of life then decided
That they were special
Being part of a grandiose scheme.

Yes, the human mammals
Came to believe
That they were much more than
A natural outgrowth
Of the organic world—
That they were placed here
for a special purpose.

They claimed that God's imprint
Could be seen all over nature.

Someone asked "Where?"
Then they got all mad.

## Dreams versus Reality

Night-dreams are not of the same category
As when one is awake, but, obviously,
The same simulation model
Of reality is employed.

In a night dream there are
No sound waves coming in,
But there can be sound,
No E/M waves coming in,
But there can be light and vision, etc.

When awake, there are both the inputs
And the representations.

# Beneath, Below, and further

In succession due does the large give way and rule
To the ever smaller, the tiny, the minuscule,
And onto the negligibly insufficient 'awol'
Of not really much of anything there at all.

Yet, it is at this bottom herefrom that the all
Of the upward progression begins its call,
And so here the answer lies, to the sprawl,
At the boundary where nature wrote its scrawl
Of existence upon the non, and back and forth,
A place not necessarily like that we think it is,
A lawless, formless realm that's ever been the quiz.

Stability, too, has decreased, downward, woefully,
Melting within our descending journey,
And so we must meet the perfect instability
Of the potentially perfect symmetry that cannot be,
For, not only is it that everything must leak
But that there can be not even one more antique
Of a controlling factor lurking about,
For of anything else we've totally run out.

Here, then, the pulsations and the throbbings
Of the energetic vacuum that must ever swing
Between being and not, ever averaging to nothing
In its rise and fall, alternating here-there, varying.

Eternity and his elemental fellow rhymes
Of Anything and Everything bide their times,
Of which they have and always had continually
All of the time of everlasting perpetuity,
And, so, then, if one waits long enough,
Which is but an instant in forever's trough,

Say, for a months of Sundays in donkey's years,
Then not only do the rarest of events come to pass,
But, eventually, so do all things possible that last.

(The last portion is in case Big Bangs,
Or any size bangs, are an even within
A larger Totality than
Just the bang and its local results.)

# There Must be the Causeless;
## Thus, that <u>is</u> the Super TOE!

The train of thought has driven us to the answer,
Of all that borne from 'near nothing' onto eternity,
Of the origin of the original disorder,
The lone dawn of our trackless radix,
Via the rails and tunnels that ever ran out:

There cannot be ever more and more
Causes beneath even more extended causes;
Therefore, intuitive or not, the causeless <u>is</u>,
Being such as what we observe it in the quantum.

Thus, cause is only of our higher realm,
As downward thence to its root emergence—
'Possibility' needed no mother but itself,
An egg burst open, born without a chicken.

The causeless bottom is the potential
Of possibility that is/was ever there.

Since it's 'defined' as an undefined chaos,
There's no problem of no initial definition had,
Since it can't have one and so it needs not any.

Things themselves become and go
Of 'virtual' potential,
Some things remaining as the rather-enduring real.
The potential is as near to simple as it gets,
Second only to the nonexistent Nothing, of course.

So, then, the potential is of no mind or 'seeing',
For that thought system can never be constituted,

As there are no more fundamentals upon more;
for, the Potential is already the ultimate basis.

Simple things ever combine, and further up,
And/or go must through phase changes,
Leading to more complex composites/forms.

Nothing, not existing at all,
And not even being able to,
But, perhaps threatening to,
Is the simplest state of all,
So, it must ever jiggle about,
Manifesting as loose 'change'.

You might say, then, that, that is exactly why
There had to be the potential for things;
Otherwise... Total Nothing, forever.

We have now reached the unexpected TOE,
One that even satisfies the ongoing trend,
for, looking down, we've always observed
The ever descending simplicity of Nature.

Now, as such, we can't really expect to find
An Ultimate Complexity sitting
Around there at the simplest point.

We didn't find Mind there;
Thus, we are ever free to be.

This causeless bottom 'fate'...
Was/is, too, a kind of 'magical' state,
for anything could have become of it.

# Two for One Paradock Ships

It seems like you can replace
Any component of a ship,
And it will still be the same ship.

So you can replace them all,
One at a time,
And it will still be the same ship.

But then you can take all the original pieces
And assemble them into a ship,
Which, too, is the same ship
With which you started.

The ship does wear down
And also undergoes alterations, from weathering,
For example, although some of its machinery
May be somewhat more immune
To the atoms straying about,
But I guess we would still call some core
Of the ship to be the same ship,
As we do with ourselves—
We even improving until a certain age,
Which makes us never quite the same
(for we can recall our differing younger selves)
And so we grow and even die little 'deaths'
From these little changes , slight amnesias,
And new growths, until the end of life,
Which is rather a much larger change,
But our parts may go on,
Perhaps even in someone else,
With total amnesia,
As may some of the ship parts
Go into other ships.

# Why Must Totality Be Causeless?

Well, if there was something prior,
Or there was something larger,
Then it's not Totality or the All;
However, we are already speaking of Totality.

Thus it is eternal and infinite, as well,
Having neither beginning nor end,
And is also unbounded in extent,
Both in the large and in the small.

The fundamental causeless itself
Can have no prior input
Or more-fundamental subparts
To make a predictive working system,
And so it must be random and thus paradoxical,
Even though what then comes out of it is the
[Arbitrary] state of the law of no laws.

# The Spread of Consciousness

As for mind (and consciousness),
Which are really just extensions of the brain,
I would expect that it is spread all over the body,
Such as even to all of the nerve spindles,
Kind of a way of actionizing without
Having to commit to any movement
(Reports still come up the spine
To the brain, of the scenarios),
And, of course,
That trying to talk about
The 'mind' being separate
From the body-brain in any way
Makes no real sense,
Except in some language shortcuts
Such as "What's on your mind?",
This being really saying about that which surfaced
In the more global/higher part of the brain
At this instant or moment
(Usually a very few things at any given instant),
From that repertoire that is your self,
This being what you have amounted to
From memories, instincts,
Experiences, learnings, and all that.

Interestingly enough, though,
The common usage of 'I' seems to refer
More to the conscious sensation,
Rather than the self part of the brain,
Which is mostly hidden in its subconscious analysis,
This analysis even being rather done and finished,
Well before the results reach our consciousness,
A few hundred milliseconds later.
This gets back to almost having to hear

What we are saying or writing
To know the full depth of what we think.

It's not that we would want
To see the neurons firing away and calculating,
For that might make little sense,
Nor even the rest of the 95% that the brain does
Without it ever reaching consciousness.

Walt Whitman had a good description
Of the body-brain-being-the-same.


## Zeno's Paradux

(This was Zeno of Kazistan)

Although a pair of ducks
Crossing a pond
Will swim in a straight line
They will not reach the other side;

However, by the time
Zeno had finished his proof
The ducks had reached the other side.

The exception was a daffy duck
Who always seemed to only get
Halfway across, ever deciding
That it would be too tiring,
And so it swam back.

Meanwhile, Daffy was passed
By a tortoise that had
Started from the other side.

## H2O Paradox

Don't go near the water
'Til you have learned to swim.

## Zero's Paradox

If it is zero degrees today
And it will be twice as cold tomorrow
What will the temperature be?

## Free Will Paradox

Only the religious believers have free will.

This is because God said
That they must have it and use it.

If if disagrees with God's will, then, well...

## Which Came First,
## The Egg or the Sperm?

The people came first?

No, Sperm, eggs don't come.

## Slip-Sliding

One foot in an empty grave,
And the other on a banana peel.
(The slippery slope of factless insanity)

# The Universal Dance

The simplest finite something
Is natural, never the less!
Being normal, as well,
Surprisingly, is motion, not rest!

The quantum jitterbugs arose, in a dance,
As they WERE ever wont to do, perchance,
Because nonexistence cannot be a reality;

These fluctuations create particles freely,
And their antiparticles, that eventually
Cancel back to zilch, or near entirely;

So, in a way, all that is 'reality'
That is somewhat enduringly
Present now is but an expression existing
Of Nonexistence's absolute instability,
Or at least of Nonexistence's impossibility.

One could say, then, that all is for naught,
(And of), as it is involved in the sense bethought
That "nought" is so perfectly unstably wrought.

And so are the fluctuation's
Simpletons emitted not so stable—
Sometimes becoming undone;
for simple things seldom
Remain unrearranged,
As they go through phase changes
And recombinations,
And so forth, as deposits,
Onto and unto the
More complex composites.

# Zippo

The one and only
Eternal, steady-state,
And infinite universe
Must sum to zero,
And is therefore
A distributed form of zero,
With no net change,
Frozen in an ultrastasis
On the largest scale.


Why is the Universe
So Large and Endless?

It is because the Planck size
And that within it is so small.

Such is the consequence
Of the smallest infinity,
And vice-versa.

Finiteness can only
Exist in-between,
At the mid-point.

## Selection

Since all things are entangled,
Or because at least that nearby events
Influence one another,
The quantum knowing-all-at-once
Of superposition takes the path most probable
(A kind of rudimentary
'Perception' or awareness),
That of the least action
To obtain the most efficient result,
This information of 'bit'
Leading to the reality of 'it',
Just as elections flow in photosynthesis
With over 95% of efficiency
In green sulfur bacteria.

I'm not saying that this
Is any instant evolution thing—
Far from it, since life took billions of years
To progress and it even went off
In thousands of doomed directions
But it did work,
Albeit with a lot of luck to boot.

The answer to why life seems to have
Some amount of know-how
To its forming could be:
If consciousness is not reducible,
Being a fundamental mechanic
Of portraying information—
It is that atoms and molecules
May have, to their small extent,
Some kind of notion of awareness
Of what's going on.

# The Sum-Thing of Zero

Matter and antimatter are always created
In equal, yet opposite amounts,
Whose electrical sum is zero

Positive and negative electric fields
Sum to a neutral universe
With a zero net electrical charge.

Energy is conserved in all interactions;
The magnitude of the universe's energy
Has zero change

Space is a collection of points,
Little bits of nothingness itself,
Which embodies a geometric zero—Null

Charge must be conserved in particle interactions;
The sum of the difference between charges is zero.

Momentum is conserved,
So the universe's net momentum
Remains constant, at zero

Nonexistence is so unstable
That it jitters back and forth,
Into and out of existence,
Producing particles in pairs,
Some of which separate far enough
Not to immediately cancel out,
Which makes for real substance
That is somewhat enduring.

From there, all is as science has it.

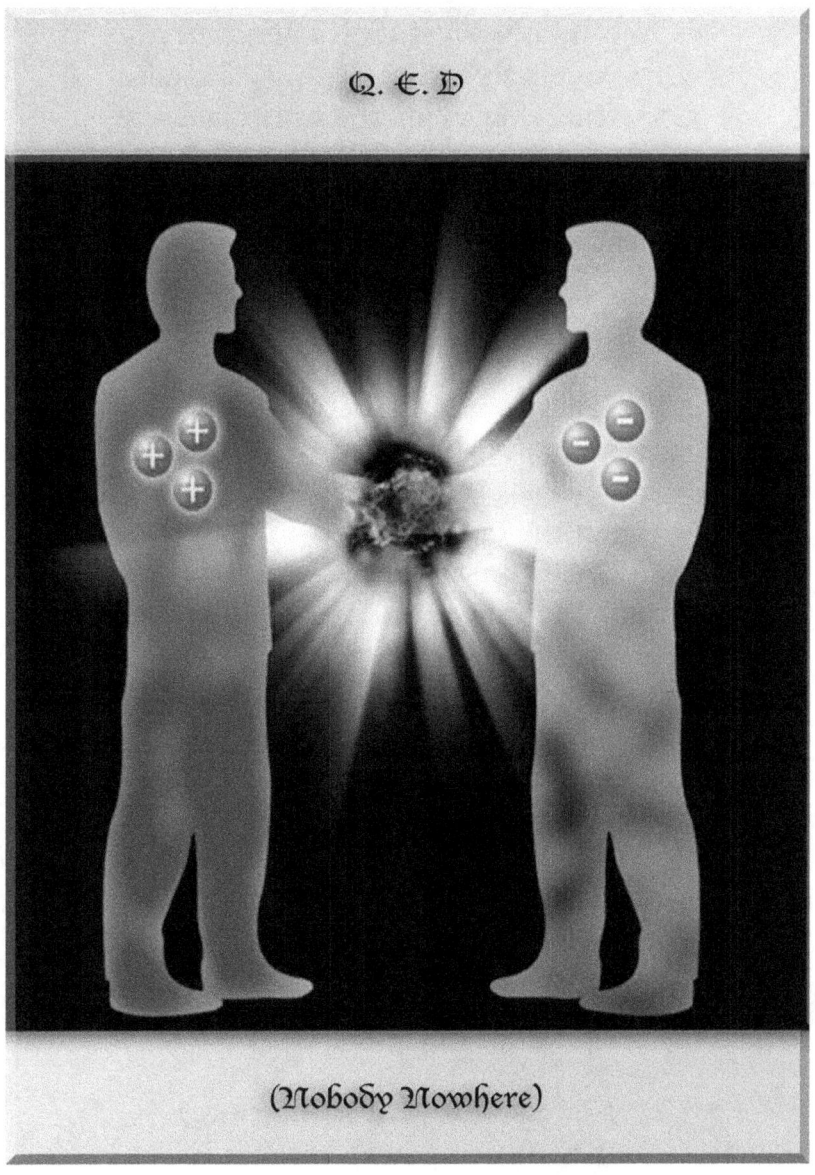

## Consciousness

When anesthesia chemicals are applied
To the brain's nerve cells
And their microtubules structure,
The ongoing process of consciousness
Of witnessing what the brain is doing
Is totally inhibited and halted.

It happens, as sure as anything, when done right.

The inescapable conclusion
Is that the process of consciousness is chemical;
There's no real decent way
To neglect or get around this.

When the anesthesia is taken off,
Then consciousness returns.
It's not that it ran and hid in the closet
And then returned when the coast was clear.

I call consciousness—
It being the brain's perception of itself—
The sixth sense.

This consciousness process extends all the way
To the nerve spindles of the body,
Allowing one to actionize
Before committing to actually moving.

It is also useful for learning,
It then being used intensely
Until the actions become more automatic.

Of course, it's greatest use

Is so we can know what we're doing,
Thinking, or feeling.

Some have latched on only
To the word 'consciousness' itself,
As if no brain or body is required to support it.

The same with the word 'number'.

Look at Kepler: first he invented God
And then he further invented upon that
That God invented numbers.

A number is just a shorthand
For a specific amount of items;
Numbers were invented by man.

(And so were Angels.)

Anyway, we have localized the phenomena.

Consciousness is a fundamental force
Like mass, space, and time,
and so it requires
No explanation—it simply arises:
Mind: it matters; matter: ever mind.

?

# Factless Insanity

## The String of Errors:

### 1.

A theory is proposed, proclaimed really,
About invisible happenings, with no evidence,
And then even further layered upon.

The schemes vary,
Ranging from All is a Dream,
Nothing is Real,
Allah Did It,
Reincarnation,
Consciousness is All (first),
Conspiracies, Astrology,
Alchemy, Tao, ETs, Numerology,
Paranormal, Ghosts, and more.

That the schemes vary is a clue
To their insanity and made-up-ness.

### 2.

Science fact is ignored.
This is because of the interest in (1).

For example, Evolution never happened
(Yet all fossils are in the right strata),
But it is that social engineers did it.

Then we may see a flip-flop:
Evolution becomes true
Because God directed it.

This betrays the interest in (1).

Or, eyes don't take in electromagnetic waves,
Nor does the brain's visual systems process data.
But, the next thing you know,
They have lapsed into
Talk of visual systems.
Insanity.

Essentially, they will dig up any old weird,
Pseudo-science stuff and hurl it against science,
Not even understanding the information.

They may even say something in favor
Of the physical material by accident,
Again not realizing, then try to disown it.

3.
They forget that (1) is just a theory
(Based on nothing, even),
And proclaim it as fact and truth.

Deceptive, unethical, insane,
And even disgusting when taught
To innocent children or to unsuspecting adults.

4.
Their only possible comeback, it seems,
When confronted with (1-3), is to get mad,
Insult, and/or claim some insult upon themselves.
Anger has no brains.

The theories of (1) are based merely
On internal neural sensation,
They not even being informed by
What science has found beneath or externally.

The viewpoints can be ignored, humored
Bashed, shown to be nil,
But the people won't care,
For they can't.

The nonsense has been grooved
In their wishing brains
By a constant firing and wiring.

5.
There are no more errors to make,
For they have made them all.


Not Here, Not All There

A no-facts-allowed, person,
Who supposedly doesn't even exist as a person,
Who holds that eyes don't take in photons,
And that blood doesn't course through arteries,
Bringing oxygen to the brain,
Which doesn't exist,
Has referenced "Factless Insanity"
As being 'emotional',
As if the cases described don't happen;
Well, he couldn't come out and say this,
For the cases do happen,
And so the only lame resource option available
Was to label it as 'emotional',
But neurotransmitters supposedly
Don't really happen, do they
For it is consciousness that flips
The switches of the now, ironically, real brain.

Some will look forever for the imagined, real,
Projector out there that does it all,
But they will never find it,
For not only is the evidence
In these case totally unconceivable,
But is not even allowed as fact,
For facts are out.

So, the best TOE Theory Characteristic
Is not that some real, super computer projector
Is sending out a simulation of all thoughts,
Things, and actions that 100% exactly matches
Just how everything would be as if it were real,
Down to the smallest particles
And their trillions of movements per second.

This real computer would have to be
Much better than that of the Sims series,
For that looks cartoonish.

So, if my fake lamp goes off
Because the fake plug falls out,
Then the simulation,
Which tracks each and every happening,
Gives the fake look of no fake light coming
From the fake lamp via fake electricity.

There are also thousands
Of soap-opera writers
Individualizing the scripts
For us fake persons.

# I Can't Seem to Get a Ticket

A day or so ago,
When I had a continuing eye infection
And couldn't wear contact lens
(And didn't have glasses),
I went out at 3 AM to buy
Some cigarettes for my sweetheart,
Who had a bad cough
(And thought that they might help, ha-ha).
I went around a curve,
Putting on my turn signal for a right turn,
But it went off, due to the curve straightening.

The policeman looked at my 60-year old self,
But still had to ask,
"Causing trouble, taking drugs or pills, etc."
He asked me to follow a moving pen with my eyes,
which I could do very well, being nearsighted.
He then asked, without bothering to look,
"Driver's license OK?" (I said "Yes".)
Then he left, after asking what I was buying,
And having a good laugh,
Saying, "Have a good night."

He never noticed or checked the car.
It wasn't even mine.
Nor did he notice my lack of lens or glasses.

It is impossible for me to get a ticket.

(Oh, and I went to the voting booth the next day
And couldn't even see who I was voting for.)

Once I was driving an old junky van,

That had a jumpy speedometer
Which didn't really work very accurately.

A policeman stopped me,
Saying that I didn't even hear his sirens
Or see his light for two miles,
Which was near true,
And that I was going over the limit.

He took a look at the rolling junk pile and said,
"Better get that speedometer checked"
And didn't give me a ticket.

It seems that I can never get a ticket.
Sometimes they note that my license is clean
And don't give me a ticket, for that reason.
I have many other stories about not getting tickets.

Yet, good speedometers are very accurate.
Its all in what you pay for

Another time,
I had a new red '84 Honda Prelude,
And had just raced down a long steep hill
When I happened to pass a police officer
Who was standing just outside his car at the bottom.

Well, he got covered with my road dust,
And so I even sped up,
Knowing that he would come after me.

When out of sight,
I pulled into a new subdivision being built,
And right on into a garage, and closed the door,
Waiting there for three hours or more,

Since I knew the police would
Have set up road blocks everywhere.

At one point I peeked out the window
And saw the police car
Driving around on the next block over.

I didn't get a ticket since they never found me.
He probably had a good radar speed indicator.

Another time, in my teens,
I had just gotten a motorcycle,
Racing it down one of those clear, long streets
That they have next to interstate highways,
When the police checked their
Good speed indicator thing,
Whatever it was in those days,
And began chasing me.

I zigzagged for a few blocks inward,
Gaining a lead, then pulled into my own garage,
Laying the motorcycle down behind a car.
I saw some reflections of flashing lights'
And heard the police car driving by, keeping going.

Oh, my license plate?
I had rotated the metal town vehicle tag
Over the last two numbers of the plate, early on.

Once I threw a lit cigarette out of a window
And it stuck into a cop's eye,
Plus it was his last day to make his ticket quota.
I didn't get a ticket.

(I made that one up.)

# Proof of the Non-Statistical

Stuff had to be; no alternative; it's here.
'Nothing' can't stay. No choice; no mystery there.

...Eternal basis of the stuff, or the stuff itself,
And probably to infinite extent (at least space is).
Nothing for a Creator to do,
For the basis of all was never created.

Is it an open question
As to whether all is completely random,
Just random statistical,
Only partly random,
Or all determined,
(Even given a random initial condition)
By some kind of perfect equation
Granting the conservation of energy law, etc.?

Yes, but we will close it.

These are even things that can be tested
By looking at the arrangements in the universe.

Also, if we can we get down to
The itsy-bitsy 'fundamental substance',
Then we can try to see if it is unbreakable,
And therefore unmakable.

A truly fundamental entity of stuff should
Have no pieces that it can be broken in to;
So, if it doesn't break...
However, without infinite power,
I guess we'll never know.

The only nagging thought would be:
How could elemental stuff be already made,
Without it ever having been made?

Yet that must also apply somewhat
If stuff comes and goes as balanced opposites,
For there is still a basis of some medium,
Even if of zero energy,
But at least it is only one basis.

So, we conclude, still, going either way,
That the necessarily ultimate and causeless basis
Must have been around forever, it being eternal,
Thus, it, itself, could have had no creation.

A heck of a lot of stuff is out there,
In unimaginable quantities,
Maybe even an infinite amount,
If there is an infinite space to hold it;,
But I'm not so worried about this one, exactly,
Since there is truly an untold amount of stuff.

Since there would have been nothing prior
To be able to decide a specific amount of stuff,
If it were really fundamental,
I'd rather think
That basic stuff probably springs up all over,
Going away, as well, everywhere, varying,
Or at least that it is infinite.

Getting rid of the 'chicken and egg' problem,
Such as there being no first thing, no first star,
No first light needed first to make matter,
No first matter needed first to make light, etc.,
Would seem to require that all be made at once.

At any rate, The Mother of all Science,
In her karmatic carriage, runs over the Godma.

The GodMother was then run over by a reindeer.

And the funny thing is, too, that life always was.
Does anyone realize this?

There could not have been a first star!

Nor any first light of photons
That split into the matter
Of an electron and a positron,
Nor any first electron and positron
That annihilated and produced a photon of light.

Nor any first electron,
Nor any electron before there was a positron.

What are the implications of all this?

That's what I like to think about.

I have only solved the last one so far:
Electrons & positrons are made at the same time.

No first life, no first kiss, no earliest history,
No last star, no last kiss...

Other matter and antimatter, too,
Besides electrons and positrons,
Like maybe quarks and antiquarks,
Produce light when they meet,
The matter converting to energy photon waves,

Their plus & minus aspects living on in neutrality,
But at some point arising again, separately,
Back as and into the matter/antimatter.

There's no memory needed for the first doings,
Since there weren't any earliest ones, plus
The history of all events repeats everywhere,
And always will, forever, and always has.

Are we getting anywhere? Yes.

What does all this mean
For local human perspectives,
Worries, and so forth?

I have some proofs showing that
The distribution of energy in space (everywhere),
At this very moment is not a statistical accident,
But is, in fact, an absolute certainty.

There are several proofs,
Each of which can each stand on its own.

(Note, that for my notation here,
Totality more or less
Equals the one and only universe
Equals space-time
Equals everything
Equals reality.)

Aside: It might also be
That all is electrically symmetric:
Being 50% matter and 50% antimatter,
Although this may not matter,
Except for the 4th proof.

#1.

## The Fixed Equation

Existence contains all the details and information
In an infinitely precise balance.
It is not a gigantic chaotic contraption
Forced to obey some general rules.
Reality is an utterly flawless system
Even though the basis itself
Is causeless, eternal, and infinite.

Every electron and photon
Has the exact momentum
It is supposed to have,
To infinitely fine resolution,
Because a photon (or an electron)
Has a history,
And the energy of which it is composed
Has an infinite history.
Thus, infinite precision.

There could not be a different Totality
Exactly the same as ours,
Right at this moment,
But with one of its photons
Displaced by one centimeter,
For that would break
Its connectivity to all else.
Photons and electrons cannot
Just randomly do any old thing;
Besides, they are not miniature first causes,
Not to mention that then everything
We use then for would also go haywire.

### #2.
### Conservation

Energy, and more, must be conserved,
And so the total amount cannot change.
A photon or an electron
Cannot do other than it must,
Else the conservation laws would fail.
There are no such 'special' universal moments,
Not even any special moments at all, as we will see.

Energy level change vanishes in the Overview
Because of this infinite precision,
For all particles must move in a precise way,
Which we will get to as more 'why' later,
As it is the basis of the conservation laws.

Universal energy distribution is fixed,
For, if it wasn't, then it would mean,
From any new information
That the prior state was incomplete.

(About Completeness)

The infinite life span of All
Provides more than enough time
For the energy and matter of which it is composed
To shift through every available internal state;
Eternity is long enough for a large
Or infinite amount of matter/energy
To express every possible spacial distribution.

All history and future occurs all of the time,
Totality containing its own history and future;
So, the universe's energy is complete.

Of course, very large distances,
Even infinite ones,
May intervene,
As to Earth's nearest moment elsewhere.


This is all showing how vast
And immutable Totality is.
It's not like we can actually go large distances;
However, nothing is ever lost;
Each and every kiss lives on, somewhere.
We are all of the indelible community of certainty.


(Infinite Eternity—
Interlude about some reminders of eternity)


The Earth is here,
Somewhere in a non-special place in Infinity,
At a non-special time,
In a small parentheses of Eternity.
No moment has any meaning more than any other,;
So, essentially, they are all the same


Eternity means no first origin
for our boundless reality,
But that is not the only example.


Matter is necessary to create light
And light is necessary to create matter.


Stellar ignition of stars also requires the presence
Of heavy and perhaps even radioactive elements,
Yet the only place where these are created
Is in a star's core.

So, since reality had no first origin,
Neither did its fundamental energy forms or stars.
The key is balance and equilibrium.
There will be no chicken and egg problem.


# 3.
Variability Inversion


The larger an object,
The less its universal variability,
Which is precisely the opposite
Of what would be expected in a statistical universe.

From a purely statistical standpoint,
An object's variability ought to increase with size—
The larger the object, the more complex;
However, although large objects
Appear to express greater diversity,
They are physically unable to achieve it,
Their size dictating their cosmic number density,
This in turn limiting their cosmic variability.

As the scale of size extends,
Cosmic variability continues to fall off.

There are fewer stellar patterns
Than atomic patterns.

There are fewer galactic patterns
Than stellar patterns.
(The total number of different galaxies
Is at least a hundred billion times less than
The number of different stars,

Since the average galaxy contains
About one hundred billion stars.)

There are fewer supercluster patterns
Than galactic patterns.

Going on up to Totality itself,
Variability continues to fall off,
Finally reaching the cessation of variability,
It looking like some kind of stringy, uniform foam.


(About Nothing)

The only possible causeless prime mover
That is both infinite and eternal is Nothing;
Else, we have infinite regress or, sillily,
The All was not the All, which cannot be.

Nothing, being the simplest state,
Must then be the most unstable state,
Even perfectly so,
It not even being able to stay as such,
Even for an instant,
Having to produce
Plus and minus material
As something like
Do the quantum vacuum fluctuations,
Emitting opposite particles
As 'virtual' pairs,
Some canceling back,
But others enduring,
In an ongoing process.
This leads to the symmetry
Of the conservation laws.

# Proof # 0
## Of the Non-Statistical Totality,
## As well as the Solution of the
## Chicken and the Egg Problem

We know that Totality could have had no origin,
This meaning that it is causelessly eternal,
Because, if not, it wouldn't be the All,
Plus that there cannot be an infinite regress
Of other, smaller entities, forming the stuff.

Systems with no origin
Are called 'temporally closed'.
Strangely, they have to be their own precursors.

Since the basis can have no cause,
We must drop the cause-and-effect at the base level,
Replacing it with a perfect equation.

It is as if no-origin systems go round and round,
Everything they need being there beforehand.
More like a perfect equation
Than any prior cause-and-effect,
Which is impossible anyway.

Reality pulsates,
In its real and structured sequence,
A field that's present throughout space immense,
Out of which all particles can condense—
Occurring where the field's extremely intense.

Particles are those bundles of inertia,
The knots in the field and fabric of space;
Yet, matter defines the structure of space...
So the Yin is in the Yang, and vice-versa!

Now we also have that matter requires light
For it to come into existence,
But that light requires matter for it to form,
Plus, that stars (of which there was no first one)
Require their own future byproducts
In order to achieve ignition.

We didn't take the implications of eternalness
Far enough, halting too soon.
It means that there is a definiteness
And a certainty as to always having
What is required, plus, it is ever repeating
By going through all of the possible events,
Even multiple times, due to eternity, plus.
Even more multiple times, due to infinity.

Think of two locked boxes,
Each of which contains the other's key.
It seems that both boxes must remain locked,
Yet, due to eternity,
Each of the yin/yang boxes can be unlocked at once.

Eternal systems are ever their own prior state,
Running like a Mobius strip
In which their present can utilize their future.
How amazing is this!

It's all of a piece,
The entire future and history of the universe
Existing all at once, inside it,
Since everything exists everywhere, many times,
And also because it can use its own products
Of the future in the past,
All of its Totality connected, post and prior,

And operating with absolute precision,
Like a clockwork orange,
It being the only way that it could be.

There is also the aspect of
The three and only three stable particles,
The electron(–), the proton(+),
And the photon ('–' and '+' as neutral),
Which hints of only two ways to make things,
Which is a part of the 'why'
Of the top-secret proof #4
That the world may dread to see.
(The Pope will cry for more than three days.)


#4
The Prime Mover

The universe is a perfect equation because
Its precision is required
for it to sum to nonexistence(!),
for the only possible infinite
And eternal prime mover is Nothing
(Else, an infinite regress).

All that we know and love
Is but a distribution of Nothing,
Such as noted in the opposite pair production
Of virtual particles,
Some of which stick around for a while.

This is all because the necessarily ultimate
And causeless basis had to have been around forever,
It being eternal, thus, it, itself,
Could have had no creation.

This alone made the Popes cry,
From the pain and injury
Of their old dogma in stone falling,
From there being no creation or Creator
Of the eternal causeless.

Adding insult to injury,
We now see that All
Is merely a balance of nothing.
Well, we knew that the TOE had to be simple,
And thus not all that interesting.

Since there would have been nothing prior to decide
About any specific amount of ever-existing stuff,
Plus, more, if it were truly fundamental,
Then it's more likely that stuff springs up all over,
Going away, as well, in time, everywhere.

Getting rid of the 'chicken and egg' problem,
Such as there being no first thing,
No first star, no light needed first to make matter,
No matter needed first to make light, etc.,
Would seem to require that any opposites
Somehow be made at once,
Or as said, always available,
Since no-origins systems are their own precursors.

The perfect symmetry of Nothing,
Combined with it not being able to stay as such,
Not even for an instant,
Due to it being the simplest state—
And thus perfectly unstable,
Is the basis for all the conservation laws.

There are only two stable matter particles,

Electrons (–) and protons(+),
And their antiparticles,
Because pair production has only two states
Able to generate separate matter particles,
These being the two canceling halves of Nothing.

The only non-matter particle, the photon,
Has neutral charge,
Being of both plus and minus combined.

Only these 3 particles can be stable,
No more, no less!

It is the summation of equal amounts
Of opposite electric charges
That nullifies existence in the overview
Of the distributed form of Nothing.

We, too, are where we must be:
At the finite midpoint
Between infinite smallness
And infinite largeness.


Conclusions for Now


There is no 'random',
Not even a probability-statistical one.
There are no first causes beyond that
Of the bottommost causeless state.

One cannot even make a truly random generator,
For, when one tries, such as in computers,
One must even keep a history
To maintain its regularity of randomness

(When designing certain
Computer instructions for rounding,
Or for other reasons).

I used a computer random number generator once
To shuffle a deck of computer cards
For the game of Hearts,
But it always went on the same
If the initial seed was put back in,
Which was at least good for replays.

It still could be that the basis of all,
Itself, is still stuff,
Stuff that was around forever,
But I like to improve on that,
For 1) What could have decided
Its amount and nature,
And 2) How could stuff be already made
Without ever having been made?
Yet, this is the only other choice,
And so it must be considered,
As being one of the two 'impossible' choices,
For, one of them must be correct,
Which should relax any notions of incredibility
When picking one or the other.


Time forever

It's easier to figure eternal 'time', maybe,
By noting that the 'sum-thing'
Always had to jiggle,
Its crests and troughs being
The positive and negative energies going around,
Even if the crests and troughs sum to zippo, overall,

For what other eternal and infinite
Prime mover could there be.

So, time is really just
This jitterbugging energy going around,
It being the energy, the charge, and the motion,
Going on forever, as ever it always did.

If it's somehow outside of time as we know it,
Then it just is, rather than not, which it can't be,
Since something is here, perhaps just because
Nothing is so perfectly unstable.

As for systems that are temporally closed,
Such as those which have no origin,
Having been around forever,
These systems are their own causes.

Now what?

Nothing.

All done.

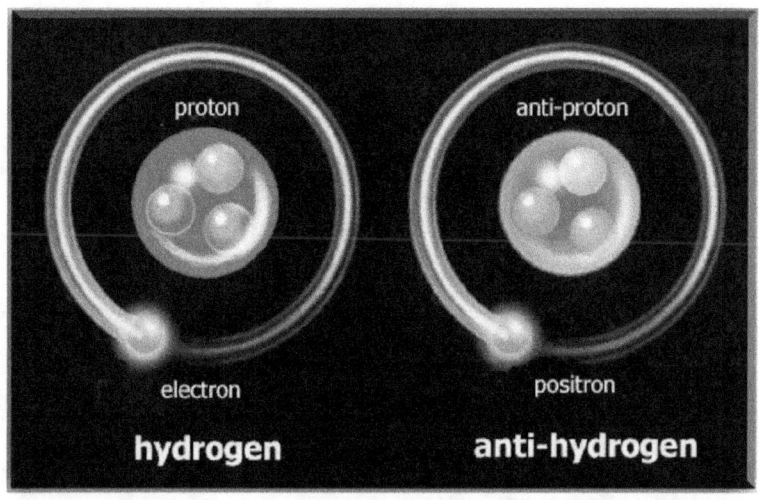

## At the Vatican in the Present

The Pope drank a lot of wine
When he heard about time,
That the basis had no creation,
It being an eternal consternation.

Then, upon finding that the basis was nothing,
He brought even more wine out for the drinking.

As he staggered up the steps to the frieze,
He noted that bad things ever come in threes.

Then it hit him, he sinking to his knees;
The non-statistical universe of the nil
Meant that there could be no free will.

After a while, he rose, somewhat gladdened,
Realizing that at least everything must happen.


## Now Here No Where

"What a day," said the Pope
To his new Camerlengo,
"But at least I can relax now,
Neglecting all the bad news of...
Well, I forgot it already.
What's your name?"

"I am Nobody Nowhere."

"Uh, oh, more bad news?"

"Yes, there are no absolute yesterdays,

Although there may be duplicates arising,
Somewhere, an almost infinite distance away,
As they have always arisen, throughout eternity."

"So, there is only now?"

"Yes, because every instant
Is immediately annihilated away
Just after it occurs!
All gone.
That is why there is only now."

"Oh, God."

"Nope, not even that,
For all is only as it must be."

The Pope looked for more wine to chug,
12% proof # 5,
But he had none left,
But, then again—thank God,
For today would soon be tomorrow;
(Hail to its obliteration).

## The Nullification of Existence

We have seen how the universe
Has three compositional dimensions (space)
And one closure dimension (fourth dimension)—
Having a finite four-dimensional hypervolume,
Its one and only one universal boundary condition
Is proportional to the product
Of Planck's constant and the speed of light
Divided by the average universal energy density,
The fourth root of the universe's hypervolume
Becoming an absolute unit of measurement
Whose length, amazingly,
Is but a fraction of a millimeter.
But, remember, the 4th dimension
Touches the 3rd everywhere,
So, it is extensive,
This tiny extent being the same everywhere—
Which is why universal constants control all,
It being the only connection
Between the micro and macro universe,
For it is the the finiteness of necessity
That exists at the midpoint
Of the universal size scale between
Infinite largeness and infinite smallness.

The four-dimensional size of the universe
Represents the quintessential quantity—
The quantity of quantity.

The only way it can exist
Within nothingness (the prime mover)
Is by being voided by some internal relationship—
A substructure allowing for
Complete and utter cancellation.

Nonexistence is the same vacant truth
On both ends of reality
And existence is everything in between.
The nonexistence of half of the universe
Must be equivalent
To the existence of the other half,
This being called existential parity.

The two halves satisfy
The completeness of the universe,
But what about its voided nature?
Some operator—a difference operator—
Has to offset the cumulative effect
Of the summation of nothing;
The largest possible difference in the universe
Is the one occurring between its two halves.

Positive and negative directions
Along the fourth dimension
Are entirely equivalent;
The only difference is
Their opposition to each other—
A polarity inherent in the fourth dimension
Required by the symmetry of totality.
That is why there is an infinite wealth of
Positive and negative electric fields
Scattered across space.

Totality's hypercube consists
Of an infinite number of layers
Of three-dimensional space
At various fourth-dimensional elevations,
This fourth-dimension not containing points
Because it represents

Their fourth-dimensional deflection—
A difference of position.

Space exists because
The sum of nothing is nothing
And the fourth-dimension exists because
The difference of nothing is nothing.

The positive and negative fields of energy
Are physical deflections of space
Along the fourth dimension.
Since all is composed of space,
Anything not strictly space is spacial distortion.

Polarity nullifies spacial magnitude.
A spacial point and its deflection
Are not separate entities.
If we could superimpose
The two halves of totality,
The result would not be space,
But nothingness.

Existence contains the exact amounts
Of positive and negative electric fields
(Spacial deflection) necessary to provide
Perfect four-dimensional symmetry.

What is fourth-dimensional,
Intrinsically polar, external to space,
And a metric for spacial distance?
Time.

Time, like space,
Is an inevitable consequence of hypervolume.
Space might constitute

The composition of reality,
But time is the cause and effect
Binding it all together.


Time is the difference of space!


Time is not a compositional dimension;
It is a difference dimension.


The hypercube has dimensions
Of quadratic distance;
However, any incomplete representation
Of this hypercube,
Such as half of unit hypervolume,
Has units of time-distance^3.
Totality is neutral and symmetric,
Whereas its internal composition
Is polar and asymmetric.


Time is the dimension that bounds,
Not extends, three-dimensional space.


The speed of light (c),
Is the underlying dimensional relationship
Between time and distance;
It provides the standard
For unbounded duration,
Much as the universes "diameter"
Provides the standard of unbounded distance.


'c' is a ratio!


$$Distance^4 / (time\text{-}distance^3) = c = distance/time$$


Why three spacial dimensions?

The singularity of nothingness
Demands existential closure,
Which in turn demands compositional parity,
Which in turn demands cubic space.

Our universe's dimensionality
Is as inevitable as its existence.

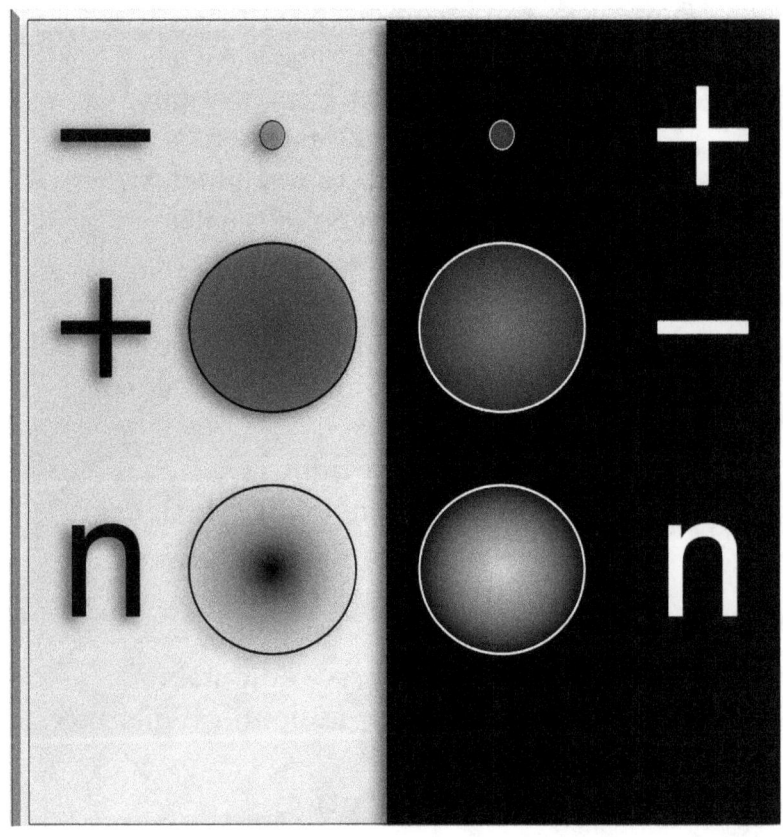

## The Spirit of the New Illumination

In the year 2031,
St. Austino was working
At the VLHC at CERN,
Looking for God's fingerprint,
When he received an invitation
From the Pope herself
To visit her in Vatican City.

All travel took place at night now,
Due to the ozone holes, and, indeed,
Many people now slept in the daytime,
Next to a fan, after taking vitamin-D.
In the evening they took light therapy.

Austino arrived in Rome within the hour,
A helicopter taking him
To the platform near St. Peter's,
Landing about 3 AM.

Austino got out and soon noted
A monument honoring the Illuminati.

Things had sure changed here over the years.

Pope Teresa—the first,
Bounded down the steps to greet Austin,
Saying, "Thank God for science
Fixing some of the ozone holes,
But I have really come to love the night.
The days are of course still too warm yet."

"Well," Austino replied,
"It will take years to replace all of the ozone,

But the plan is working,
And I'll thank God. too, if I ever find her."

"I haven't found even one
Of her fingerprints, Austin.

"And I have found none myself."

"Agreement at last."

"Well, Pope, it was a very beautiful
And glorious wish, just the same."

"Yes, for it brings much happiness to Sapiens,
But then again,
So do other wrong things such as drugs."

"True, as qualified,
And it's even that natural selection
May have put the wishes there."

"Darwin's idea was the best idea
That anyone ever had."

"True, Ms. Pope Teresa;
Well, shall we let joy and innocence prevail?"

"Yes, perhaps,
For at least one more generation.
Attendance is falling."

"The Very Large Hadron Collider
Was the last hope."

"Thanks for looking, St. Austino.

We know that you were hoping not to find God,
But your actual research was fair and unbiased.
You left no rock untorn but St. Peter's."

"Thanks, Ms. Pope. I tried."

"We are surely on our own
In this universe now, old Austino."

"One can be alone but never lonely,
for we have our lives."

"Yes, that's always where it was at, wasn't it?"

"True. And now we know that we are truly free."

"It is a liberation, really."

"We can now do good just for the sake of good."

"That's what it's all about about here,
As ever for many others,
Like those at ToeQuest."

"And science has neutralized
The WMDs of the religious fundamentalist
nations."

"Thank Einstein."

"I see that everyone is up and about."

"We all work at night now."

"I see that the celibacy rule was lifted."

"True, Austin, and so now the Pope
Can even go out on a date."

"Really. Shall we?"

"Yes. To the movies?
Angels & Demons Part 7 just came out."

"Well, is the Pope female?"

"Of course I am, let's go.
'7' is a lucky number."

"Yes, there were 7 proofs of mine."

"I am named after St. Teresa of the ecstasy."

"Oh, my, my.
I saw Bernini's sculpture
Of her with the angel.
Bernini was an illuminatus."

"True, a great guy. Care for a smoke?"

"Don't mind if I do,
for they no longer have harmful additives."

"And they still aid concentration,
but now even better.
I used to sneak them
When I was a Cardinal."

"Thank science."

She, the Pope, lit one up
And handed it over to Austin,
Then lit one for herself.
Smoke clouds soon rose unto the sky.

"There were no commandments against smoking,
Austino, and, as you know,
We do love wine, as well."

"Smoke is the spirit of the Holy Ghost."

"Ha. That's a good one!"

"Have any wine around, Pope?"

"Sure, here you go."

"You seem strangely familiar,
Miss Teresa Pope, the Very First.
What is your given name?"

"Melanie."

"Well, we meet, at last, my dear,
Under starry skies."

"Yes, finally.
And now science has doubled our life spans, Austin,
So we are now only halfway through."

"Yes, Popesie,
And now that our consciousnesses have merged,
We can really enjoy life to the fullest."

"Yes, all the prep work is finally done"

"Hey, who's that monk?"

"That's no monk;
It's Professor Pat going over to the archives
To read some fine and tiny print."

"What! He never ever even read
Halfway through my posts—
And I even used size 3 font
And put many spacing lines."

"Profpat has come a long way."

"Hey, who's that guy with the long beard?"

"That's Graybeard.
He's teaching evolution to our clergy.
Yet another big missing link has been found."

"And that lady on that fine brown talking horse?"

"LabelWench is our prime diplomatic liaison
To the scientific community,
Since she taught Sunday school
Once upon a time, and learned science from Lloyd.
She teaches us how to work at night, too.
Also, her horse, Caramel,
Speaks to the animals in their own language,
Even in cricket-ese."

"Holy Moly cripes. And who's that speedy guy?"

"That's TimeParticle.
He is the chief of all our humanitarian efforts.

And he never runs out of time.
He's also one of our resident poets,
Along with young Mohan."

"Wow! So, all this still goes on
Without there being a God."

"Yes, for if there was a God,
She would have wanted it this way.
But few are for the vengeful God of Old, a myth."

"True, plus evolution put
The spirit of this into some."

"Yes, we are naturally
Supernaturally superstitious."

"And there is still Jesus to follow."

"Yes, he was a fine sapiens
And was very much ahead of his time.
We don't need his father."

"Who's that half-invisible guy over there?"

"Nobody."

"C'mon, now;
Where does he live, here or there."

"Nowhere."

"What! Is he is the CIA or something?"

"No, Ninja Empire.

Nobody Nowhere is becoming real,
For the moment."

"And who's that guy
With all the digital equipment?"

"Oh, that's analog.
He converted, but he kept his old name."

"Who's that in the big Green Bug suit?"

"That's GreenBug;
He looks after the health
Of our environment."

"And the lady in the white coat?"

"Ms. Lesley Key is the head
Of WorldWide Health;
She is here is vaccinate us
Against the flea flu?"

"The flea flew?"

"No, flea's jump;
The people flee,
As from the bird flu."

"The bird flew?
And the swine flu?"

"Yes, but pigs can't fly."

"You're a funny Pope,
But a fitting one for these new times."

"Yes, for when the karma ran over the dogma,
They had to meet halfway;
I was the happy medium."

Who's that guy drawing circles
Crashing into each other?"

"That's Bogie, one of our smartest,
Working on arenas in space."

"But he's here, on the ground."

"He doesn't go on field trips like Dip does."

"Who that guy with fractal hands
Vibrating in and out of
Their most likely places?"

"That's Steve; he's superimposed a bit."

"And who's that guy who
Looks the same all over?"

"MJA."

"What's that sign about
The last of the bloodline lecture?"

"Tarina's coming here tomorrow to speak,
For she and her children are the last of
The blood line of Jesus and Mary Magdelane."

"Holy Christ!"

"You can't say that here."

"I mean, she's arriving!
We might get in trouble for going on a date!"

"Nah, for she has proclaimed
That it is ever virtuous to share."

"Whew!"

"Who's that guy trying
To bum a smoke over there?"

"That's Graham.
He's here to train us how to
Levitate up to the new Magno City."

"Holy Cow! So much progress.
And I hear that ToeQuest
Is now the #1 web site;
I bet Robert is really busy with that now."

"Nah, he's not busy,
For he is very talented."

"See that guy playing video games over there?"

"Well, I'll be darned! That's Meem.
See, Austin, I only said 'darn', not 'damn'."

"Oops."

Come on in, Austino;
I'll get out of the habit."

"Smoking?"

"No these flowing robes.
Then we'll take the old catacomb."

"Hey, who's that restoring the statues
To their full anatomy?"

"It's my mirror 'Melanie'.
She's my Camerlengo—my assistant."

"She really exists?"

"No, but yes."

"And who that handing her the pieces
In exchange for the fig leaves removed?"

"Racecar."

"Holy mother of Jesus;
It's like old times here.
Aren't they, um,
fabricating some extra reality."

"Perhaps, for they have to put in some filler
Where it was broken off."

"Got a job for me?"

"Want to oversee the naked art museum?"

"Sure."

...

St. Austino and the She-Pope
Soon passed through the tunnel,
Emerging into Galileo's old castle fair,
The Castele Sant'Angelo,
Then walked across
The lovely Bridge of the Angels,
Arm in arm, spirit in spirit,
To view the fabulous holographic film,
With its in-the-head-sound, odour-vision,
Air-taste, and vibrating seats.


Nobody Home

(Back in 2010),
The Pope happened to remember
The scientific revelations that had torn asunder
The rock upon which the dogma's thunder
Had been carved in stone as rendered.

So he called upon his camerlengo once again
To speak some more about now and then.

"So, then if there's no yesterday, for sure,
Then at least there is the future."

"Sir Pope, there is no future either."

"What!"

"Everything already happened, as one,
All at once, in no time done."

"It's all gone?"

'Yes, and even the present that is and was
Is but what ancient history does."

"Then what's all this?" said the Pope's nose,
Pointing around, and out of the window.

"It's just the slow motion broadcast
Of all that happened so fast."

"How come this tape-delay?"

"The speed of light,
As fast as it is,
Slowed it down."

"So, it's all set in stone,
With no alternate endings grown?"

"Yes, it's invariant, just as so,
But we get to enjoy the show."

"Another proof?"

"Yes, 100% proof—
A very intoxicating truth."

The Pope picked up his wine bottle...
And threw it out of the window.

(He had forgotten to open the window first.)

# The Equations of Reality

Infinite Smallness x Infinite Largeness
equals
Finite Unity


Past's Destruction
—> Now —>
Future's Construction


(s)

I
n
f
i
(–) Eternities (+)
i
t
i
e
s

(L)

S
p
(–) Polarity (+)
c
e
s

# The One
(solidity)

and

# The None
(vacuity)

are

Impossible,

thus

# The

In-between

zero Sumthing

of

Plus
&
Minus

The singularity of nothingness
Demands existential closure,
Which in turn demands compositional parity,
Which in turn demands cubic space.

Our universe's dimensionality
Is as inevitable as its existence.

## from
## A to Z

"Hi Aleph," said Omega.

"Just call me 'Alpha',
I have converted.

"Hey, Alpha, you know,
Sometimes things just happen,
Little by little,
Unrigged by conspiracies."

"But, Omega, I need someone to blame."

"For what?"

"Well, for what people do,
Plus I have to work,
And suffer under a system,
Ever getting insulted."

"That's not what makes you
Sad and grumpy;
You do."

"But I had to put out thousands,
Complaining for changes,
Looking everywhere for evil.
Will that do anything?"

"No, nothing,
For all is of a balance.
Nature always wins!"

"Thanks, Omega.
Where's Zed?"

"Oh, he's in Australia, on vacation."

"So, what's existence all about?"

"We reside as gleams in one of those
Shimmering paths of everything—
In our own lustrous, clayed pot of shining gold,
As the glitter within its sparkling rainbow.

We twinkle as real and glimmering rainbows,
The stable/virtual the same glint as the real—
The glowing differentiation of the balance."

"Hey, cool, I think I'll start living—
It's the most important part of existing!"

(The alphabet was now complete.)

# Location, Location, Location

Given that there is life on Earth,
Our solar system had to be
In the outer-more portion of the galaxy,
For, if not, we could have never been;
And, in fact, it is on a spiral arm.

Reality's building blocks are quite ancient,
But have been reshuffled by galactic cores.

All galaxies are vortexes similar to ours,
Very little material within appearing
Older than the galactic transit time.
This is why the entire universe
Appears to have a finite age,
And why our local neighborhood
Appears to be about 10 billion years old.

Even globular clusters seem
Older than the galaxy they orbit,
For their trajectories lie outside
The galactic vortex,
And so they are not pulled into
The recycling engine
With the same regularity
As its disk material.

It took about 10 billion years
For our solar system to reach
Its current distance from the galactic rim.
How long until it falls into the core?

The time it takes to fall through
The luminous portion of our galaxy's disk

Is about 16 million years,
Which is called the galactic transit time.

Any star born on the Milky Way's rim
With a mass smaller than that of our sun
Will still be burning when it
Falls into our galaxy's core.

Indeed, our own sun has enough fuel
To burn for another 4.5 billion years.

At its current rate of descent,
Our solar system will be in
The Milky Way's core region
In less than 4 billion years,
To be consumed by the voracious beast
Of the galaxy's Black Hole,
Perhaps, while it is still burning
So mark your calendars.

— The Infernal Regions —

Hellholes hurl thousand light-year jets of fear,
In Centaurus, cross'd the galactic sphere,
Supermassive darkling beasts devour all...
Abandon hope, all ye who enter here.

There are two unforgiving time constraints
For the evolution and long term survival
Of life in the universe.

It must advance from bacteria
To full scale space transport

Before its sun fails
Or falls into the core of the galaxy,
Whatever comes first.

Earth has already had a total run
Of about 4.5 billion years
Since our sun's ignition.

If Sol had started burning
Any closer than about
25 thousand light years
from the Milky Way's core,
Our civilization simply would not exist.

All this we could have expected,
for we are here.

## Black Holes

I forgot who said that there may be black holes
As particle cores, but, it resonated,
Although these thoughts are only if that is so.


A black hole is really a hole in space,
A discontinuity, a singularity;
It could even be considered as
Not part of our universe or external to our universe
(here there is literally nothing).


So, then, a particle core could be called "hollow",
A hole in space.
This then could be the reason
Why they are impenetrable
And non compressible past a certain point.


As fast as we can blast particles at each other,
They never pass through each other.
And there is also the electron degeneracy pressure
That limits the compression size of a white dwarf.


It is as if a particle core boundary
(Since nothing inside)
Has infinite energy,
And so particles can never
Occupy the same space.


Yet, when matter and antimatter collide,
These insurmountable cores go away in a jiffy,
As if there was nothing to them.


Who cares, for astrological connections of stars are
Really alive beings who give us good and bad days.

# Fire and Ice

A black hole's interior is composed
Of a degenerate neutron super fluid
With a gravitational potential
Strong enough to lower its own density
Through the hyper-extension of particles cores.

The greater the gravitational expansion,
The lower the  nuclear potential.

Inter-nucleon binding energy
Is reduced as well.

A galaxy's core doesn't just burn nuclei,
It first forces them apart
With its gravitational potential.

The universe is ever about balance.
If nuclei are formed in heat
They must be dissolved in relative cold.

Heat would be an inefficient waste product
For the galactic engine.
It is an environment where electrical energy
Is applied to degenerate matter,
Producing hydrogen
AND virtually no radiant energy.

The gravitational expansion required
To disassociate nuclei
At relatively low temperature
Is why galaxies require massive black holes.

There is no where else to do it.

Universal equilibrium
Demands the existence of black holes.

It is an unavoidable consequence
Of cosmic renewal,
A forgone conclusion.

Electrical current heats the galactic core,
While the absorption of compound nuclei cools it.

# Absolute Space
# And the Speed of Light

The Lorentz derivation left
Absolute space behind,
Given the assumption
That measurements
Are limited to 3 dimensions.

Relativity provides computational capability,
But does not give us access
To the underlying reality,
Yet, there is a reason that
The speed of light
Is the universe's maximum speed,
And why the internal change of a system
Approaching the speed of light slows to zero.

The speed of light is not merely
A universal constant,
But is the linear,
Dimensional relationship
Between space and time.

There can be only one such relationship,
And thus there is only one speed
For the propagation of energy through space.

Object moving at speeds less than 'c'
Are amalgamations of
Static (particles cores)
And dynamic (photon/kinetic) energy.

The observed "time dilation"
In object moving through absolute space

Is solely a product of
The universe's photonic speed limit.


A moving object is composed of
Static (rest) and dynamic (propagating) energy.


Its net motion through absolute space
COnstitutes the external manifestation
Of its dynamic energy, herein called $V$.


Any motion of its dynamic constituents
Normal to $V$ has no measurable effect
On the net motion of the entire object,
This motion being called $Vi$.


Since 'c' is the metric of
Energy's propagation through absolute space,
And since the external and internal components
Of an object's dynamic energy
Are perpendicular to each other,
Their relationship to each other
Can be depicted as a triangle.


height: $Vi$, length: $V$, diagonal: 'c'


The speed of an object's internal motion
Defines the rate
Of the internal changes it experiences,
Such as the decay period
Of an unstable relativistic particle.


An object's internal motion
Is related to its external motion by


$$Vi^2 = c^2 - V^2$$

The speed of light, 'c',
Is the metric of all change;
A motion of zero represent no change;
A motion of 'c' is maximum change.

The normalized rate
Of an object's internal change
Is the ratio
Of its internal components
To the speed of light.

$$R_i = V_i/c$$

When $V_i = 0$
Then $R_i$, the rate of internal change,
Goes to zero.

The relative length of a unit time interval
In this system is the magnitude
Of its "time dilation", $TD$,
This being the inverse of
The rate of internal change, $R_i$.

$$TD = T/T_0 = 1/R_i = c/V_i,$$

where $T_0$ is the length
Of a time interval at rest ($V=0$).

If internal change is reduced
To half of normal, for instance,
An internal event takes
Twice as long to occur.

Substituting and solving, we get

$$T/T_0 = TD =$$

$$1/[\text{square roof of}(1 - V^2/c^2)]$$

Which is the same "time dilation"
Given by the Lorentz transformation,
Yet, it is based solely
On the real, physical limitations
Of a moving system,
No reconfiguration of
Space and time being necessary.

As the external motion goes to 'c',
The rate of internal motion goes to zero.

Velocity addition can also express 'c',
By performing two
Successive Lorentz transformations,
For a V of $V_1$ and $V_2$,
This being the addition of ratios,
Not magnitudes, giving

$$V = (V_1 + V_2) / [1 + (V_1 * V_2/c^2)] =$$

$$(c + V_2 / c + V_2) = c$$

Thus, the speed of light
Appears the same way in any direction
In any moving reference frame.

Velocities could not add
Exactly like distances unless 'c' = infinity,
There then being no upper limit
To their magnitudes.

So, vector addition could not be used,
Due to the dimensional relationship
Between time and space,
Although it could be an approximation
For adding velocities.

## Just the Natural

Positing the supernatural is just a larger question
Because it doesn't really explain anything;
It just pushes the answer off.

Incomplete answers are invariably wrong,
For answers must be complete.

The proofs of the self-contradiction
Of the supernatural remain,
As well as that we find only
The natural everywhere,
The exact place that the supernatural
Is supposed to be, theistically speaking.

It is really just a pronouncement
To say that the supernatural can be,
One born of a wish.

Zero proofs of the supernatural have been found,
While there are a zillion proofs of the natural.

The error is then compounded by preaching it,
And that is the real problem.
Why can't they say that it's just a theory?
Because then fewer would listen.

Science proof is not an internal view
Like the supernatural is,
But is there for all to observe;
However, they may not want to,
For then the wish begins to wither.
Many will go through all kinds of contortions
And distortions to avoid this.

## Large and Small,
## Endless and Infinite

The one and only
Eternal, steady-state,
And infinite universe
Must sum to zero,
And is therefore
A distributed form of zero,
With no net change,
Frozen in an ultrastasis
On the largest scale.

Why is the universe
So large and endless?

It is because the Planck size
And that within it
Is so small and bottomless.

## The Stars

Stars have been burning hydrogen forever
And so they require an endless
And renewable supply.

So, the energy in the light released
By the fusion of compound nuclei
Must eventually be used
To break them apart.

## Emotions and feelings

Emotions are actions accompanied by
Ideas and certain modes of thinking,
While feelings, from emotions,
Are mostly perceptions
Of what our bodies do
During the emoting,
Along with perception of our
Own state of mind during
That same period of time.

So it is, that, as far as the body is concerned,
That feelings are images of actions
Rather than the actions themselves.

Emotions are complex,
Largely automated programs
Of actions concocted by
Evolution that are carried
Out in our bodies,
Such facial expressions, postures,
Changes in organs, and changes
In internal settings and environment.

Neither love not hate will be of any help,
In the knowledge of the TOE,
But even of some hindrance
In the full absorbence of
The meaning inherent
In fact or theory.

# Hyper Volume =

(ħc/Density) =
(E, w(wavelength) ) / D)

So we see that the dimensional units of ħc/D are:

(energy*time * distance/time) / (energy/distance^3)

Giving

(energy*distance) * (distance^3/energy) =
distance^4

(not even having to deal
With the units of energy for the moment!)

Photons are the encapsulation of time by space;
Particle fields are the
Encapsulation of space by time.

Just as Planck's constant is
The four-dimensional quantization of photons,
Elementary charge is the four-dimensional
Quantization of particles.

# Decisions

Emotional decisions can make one happy,
In everyday life as lived,
Even for such as what the TOE should be,
But, there, they can get in
The way of the light of truth.

# TOE Discovered

'Nothing' never sleeps,
But is ever up to something.
Motion never ceases;
for there is no ceasing center.

No uncertain quantum property
Can ever be zero,
for zero is a precise amount.

This road from 'nothing' to something
Goes in both directions, in the zpe.

The so-called 'vacuum' is creative.

The field fluctuates
This way and that,
But, on average,
The net energy is 'zero'.

fields can never go away,
As they're part of the
Structure of the vacuum.

When in their quietest possible state
They are the vacuum.

# The 'Maker' of All

We have to face that the fundamental substance
Is here, existing; so it must be made of something;
Yet, there is no material available to make it of.

So, then, to achieve the Toe's completion,
We must say of what it is made, and why—
Or end up with an incomplete structure.

Its only possible constitution is to be
A balance and distribution of nothing at all—
The only infinite, eternal state
Of the perfect symmetry
Of the perfect equation
That provides the balance of opposites
Everywhere that we look:
That all must sum to zero,
Although it never does so,
For that state is too precise.

There's only one way to have the speed of light,
It being the dimensional ratio of space to time;
The plus and minus charge polarity,
Matter and antimatter,
The only two stable particles—
The electron and the proton,
The one workable set of electron/proton sizes,
One way to regenerate via recycling,
Only one way of quantization of photons,
And only one way of quantization of particles.

All is as it must be,
As the only way that it can be,
Thus making it the one and only TOE.

# The Basis

We can not ever just assign more and more
Downwards and infinite regresses of causes
For the fundamental eternal causeless,
As funny as this evens sounds,
Either as scientists of the totally natural,
Or, as believers in the supernatural,
They stopping at an even larger cause,
Not a lessor one, which then should be investigated
With at least the same vigor,
But then is never done, of course,
Halting at the word being that
Which explains everything perfectly.

So, for the most underlying basis,
We must throw the notion of cause and effect
Right out of the window,
Being careful to first open the window.

Cause and effect is dear to human mammals,
Since it holds, all the way up,
After the causeless basis,
And so even when it appears that
It has been thrown out,
The often words return again,
To that cause and effect,
Such as 'design' and 'designer',
And other causes.

Now one must obtain a catapult
And really heave the notion
Really far out of the window.

What is its replacement?

It can only be an equation,
And a perfect one at that, the only one,
If that is any consolation,
Which it should be, since it makes us free.

One could also just leave it alone as "don't know",
Which also makes us just as free to be.

Freedom only disappears if one goes
To the supernatural, which isn't here,
Being proposed as 'out there',
And then goes further to say
What it wants from us, and why,
And all that extended jazz.

# Why 1/137
## For the Fine-Structure Constant
## And not 1/2

The size of the one and only universe,
This being its only boundary condition,
Is that of a 4D finite hypercube
Of quadric space made of infinite 3D spaces
Sitting atop one another,
Governs the quantization of energy forms,
Taking the form of Planck's constant for light,
And being the reason for elementary particles
Having the same magnitude of unit charge.

Charge (q) is responsible for the Coulomb force:

$$Fq = q^2 / 4*pi*eo*r^2$$

Where eo is a constant known
As the permitivity of free space.

The ratio of the Coulomb strength
To Planck's constant (hc)
Is called the fine structure constant, 1/137.

One might think that it should be 1/2
Since photons are complete,
Having an inherent positive and negative
That sums to neutrality,
The polarity only showing forth
When an electron and a positron are produced,
Whereas a matter particle
Is always an incomplete form having either
A positive or a negative charge.

The reason this is not the case
Is because, unlike Plank's constant,
The Coulomb's force strength
Is not a direct assessment of unit polarity volume,
For it is but a byproduct of unit polar volume
Just as electrostatic potential energy
Is a byproduct of rest energy.

A photon's hypervolume is geometrically closed,
While a matter particle's unit polar volume is open;
So, a matter particle's net polarity
Prevents the spacial encapsulation
Of its 4th dimensional elevation of deflection,
It thus being a distributional boundary condition,
A family of relationships,

time = unit polar volume of charge / volume.

Photons are the encapsulation of time by space,
While matter particle fields are
The encapsulation of space by time;
Thus their radically different properties.

Space is continuous and causal.
The reason why a matter particle's
Central deflection of elevation in time
Diminishes with distance is because
External deflection is governed by
A dimensional relationship
Between time and space
In 4D space-time and this relationship
Is a constant.
Spacial deflection attenuates
Due to increasing volume because
Spatial volume causes this attenuation.

Herein lies the meaning of charge.
The Coulomb's force increases
With the amount of external deflection,
But it is not simply a case
Of a maximum deflection somewhere in space,
But is a combination of deflection and volume,
And this determines the deflection
At any given range.

A matter particle distribution
Consists of an infinite number
Of concentric spherical surfaces
With 4th-dimensional elevation
Scaled by their volume and unit polar volume.

Matter particles are radically symmetrical,
So the volume element of their
field distribution is a sphere.

t = 3*(unit polar volume of charge) / 4*pi*r^3

External deflection decreases
As the cube of the distance
From a particle's center,
Whereas 3-dimensional deflections
Decrease as the square of distance,
Attenuated by increasing area,
4th-dimensional deflections decrease
As the cube of distance,
Attenuated by increasing volume.

Unit polar volume is ħc/4pi, so substituting,
We can get a matter particle's
field's actual displacement,

9.4(10^‑18) / r^3, in meters,

And so, since a proton's electrostatic field
Is active at a radius as close as 1 fermi (10^‑15)
From its energy center,
Yields an external deflection
Of a trillion light years (9.4(10^27) m).

Deflections of this magnitude are possible
Because although 'c' is energy's
Speed limit through space,
There is no limit to
Space's speed through time.